Modeling and Simulation of Protein Folding

vorgelegt von
Anna Shumilina

Vom Fachbereich Mathematik der Universität Kaiserslautern
zur Verleihung des akademischen Grades
Doktor der Naturwissenschaften
(*Doctor rerum naturalium, Dr. rer. nat.*)

genehmigte Dissertation

1. Gutachter: Prof. Dr. Tobias Damm,
2. Gutachter: Prof. Dr. Arnold Neumaier.

Datum der Disputation: 28.03.2011

D 386

Bibliografische Information der Deutschen Nationalbibliothek
Die Deutsche Nationalbibliothek verzeichnet diese Publikation in der Deutschen
Nationalbibliografie; detaillierte bibliografische Daten sind im Internet über
http://dnb.d-nb.de abrufbar.
    1. Aufl. - Göttingen: Cuvillier, 2011
    Zugl.: Kaiserslautern, Univ., Diss., 2011

    978-3-86955-798-4

© CUVILLIER VERLAG, Göttingen 2011
    Nonnenstieg 8, 37075 Göttingen
    Telefon: 0551-54724-0
    Telefax: 0551-54724-21
    www.cuvillier.de

To the memory of my mama,
Dr. Elena V. Shumilina

Abstract

The subject of this work is the development of a mathematical model for intracellular protein folding and the implementation of this model in the form of a simulation software. During the elaboration of the model special attention was given to the factors that may be determinant for physiological folding pathways. Thus, findings from biochemistry and molecular biology indicate that protein folding in a cell starts cotranslationally from certain initial conformations. Besides, the interaction with the surrounding solvent is crucial for the acquisition of the final structure.

A general expression connecting interatomic distances and dihedral angles is derived, which has allowed the formulation of the model in the space of molecular torsion angles. Twisting forces are computed analytically and utilized for the improvement of computational efficiency during energy minimization in the space of torsion angles. Besides, equations for dynamics in the space of torsion angles are derived and a conclusion related to folding pathways is drawn.

Algorithms are developed, which permit the generation of appropriate initial atomic coordinates for amino acids. Care is taken about the correct chirality of amino acids. During the simulation, the polypeptide chain is folded subsequently, as when it emerges from a ribosome during the protein synthesis. Transitions with an energy increase are allowed only to a limited extent. An attachment of a new residue is performed in a way that the formed peptide group is disposed in the *trans* conformation, which prevails significantly in native proteins.

Beside the electrostatic and van der Waals interactions, the proposed model incorporates hydrogen and disulfide bonding, solvation effects, and dielectric screening at the protein surface. The hydration is modeled without inclusion of water molecules into the simulation. Instead, the number of water molecules that can directly contact each atom is estimated with the help of a solvation grid on the atom surface. This information is used for evaluation of solvation energies, as well as for the modeling of the electrostatic screening. Solvation grids rotate randomly before each energy evaluation, giving raise to stochastic contributions from the side of the solvent.

The developed software, named SiViProF, apart from its simulation functions, also performs visualization of different kinds and contains tools for the exploration of energy surfaces for protein fragments.

CONTENTS

PREFACE

OUTLINE OF THE WORK

Protein structure prediction is a fundamental problem of biochemistry and one of the main challenges in bioinformatics. This research topic is highly interdisciplinary, since beside biochemical background it may require knowledge of physical chemistry, cell and molecular biology, computer science, various areas of physics, including but not limited to thermodynamics, statistical mechanics, and electrostatics. It can take advantage of solution methods from different fields of mathematics, starting with optimization and combinatorics, and extending to stiff partial or stochastic differential equations.

The modeling approach proposed in this work is dictated by a sincere wish to bypass weaknesses of classical methods for protein structure prediction, observed from the viewpoint of molecular biology and biochemistry, and at the same time to facilitate the problem solution using specific knowledge from cell biology, physical chemistry, informatics and particularly mathematics.

The first two chapters are generally devoted to a literature review. In Chapter 1, the background material from chemistry and physics, as well as from cell and molecular biology, which is necessary for understanding, rationalization and implementation of the proposed modeling approach, is collected. However, the contribution of the author in this chapter should not be reduced to concise formulation of known facts, also including recently discovered ones: the review is supplemented by computations and illustrations resulting from this work, in places where it was required for a better explanation of the described material or considered appropriate in the context of this chapter.

For example, the relation between the acid dissociation constants and the probabilities of amino acid residues to be ionized, as well as the connection between the probabilities and the Gibbs free energies of ionization, were mentioned in [1] (p. 123) without a proof. Other available sources didn't contain the proofs or references to the proofs either. Since this information is quite essential for determination of structure and charges of amino acid residues, these relations were also derived independently in this thesis and described in Section 1.4.

Similarly, the proofs in Subsection 1.7.2 related to the interaction of charged groups with solvent are independent contributions, although speculations similar to some of those described in that subsection could have been published in earlier physical chemistry works not considered here.

Further, the maps of energy surfaces, depicted in Figures 1.28-1.30, are generated by SIVIPROF, the software developed in course of this work. The idea of such maps, together with some related conclusions, is described in [1] and accompanied by similar illustrations, but in a schematic form.

Besides, all the other pictures, including but not limited to molecular images, were created solely for this thesis and not adopted from other sources. To produce Figure 1.38 (a-f), which depicts ribosomes and ribosomal cuts, a special module for SIVIPROF was developed, which enables creation of large non-protein molecules and molecular complexes based on atomic coordinates from RCSB Protein Data Bank records. Additionally, in order to find the intraribosomal tunnel, SIVIPROF was supplemented by a view manager that supports precise control of the viewpoint and of the section plane position, as well as an elimination of the macromolecular parts that are cut away by the plane from the sight. Adjustable black fog was added to increase the 3D effect. The module for visualization of atomic orbitals (see Figures 1.8-1.13) was intended for future visualizations of quantum mechanical computations.

Apart from the outline of the problem and its principal background, Chapter 1 brings the following two messages, which are crucial for selection and development of the solution method for this work. First, protein folding in a cell begins from a certain initial conformation, which may be essential for the acquisition of the functional final structure. Besides, the folding starts already during the protein synthesis, and it is likely that the ribosome contributes to this process. The native form does not have to be the global minimum of energy, but must be sufficiently stable and almost surely achievable within a normal folding time at physiological conditions. Second, the folding is largely driven by the interaction with the surrounding solvent. Thus, the charged atom groups tend to stay in contact with water, where they have lower energy, while the hydrophobic residues abandon the protein surface. Nearly the opposite effect could be observed in vacuum, where charges of opposite signs are attracted to each other stronger. These facts are rather well known in molecular biology. However, often they are partially or completely neglected when it comes to solving protein folding problem.

An overview of generally known approaches for protein structure prediction is given in Chapter 2. Some drawbacks and disputable questions associated with these approaches are also described there. Thus, molecular dynamics and global minimization methods applied for solution of this problem are usually based on the assumption that a protein can fold into its native form starting from any conformation. Apart from that, the global minimization approach requires that the native structure corresponds to the global minimum of energy. Besides, often the interaction with surrounding solvent is not taken into account to a sufficient extent. On the other hand, the developments of *ab initio* methods, and particularly the related model parameters, were of great importance for this work. Additionally, the information extracted by knowledge-based methods can be useful for understanding of the amino acid properties and for further model improvement.

Molecular mechanics force fields, also briefly described in Chapter 2, have constituted the basis for the proposed model. The procedure for computation of atomic partial charges, which is discussed in Subsection 2.2.3, was utilized in SIVIPROF. The specific mathematical notations, used in Chapter 2 and later on, are largely non-conventional. Although the notations found in the literature may look more simple, they are often subject to incompleteness or inconsistency. Therefore, new notations were introduced in this work, and for the sake of integrity and uniformity some of them are described and utilized already in this chapter.

The last two chapters are solely devoted to the results of the current work. The modeling approach and the theory developed in relation to it are described in Chapter 3. A general outline of the implementation, simulation results and their discussion are presented in Chapter 4.

At the beginning of Chapter 3, some analysis of the relations between the interatomic distances and the dihedral angles along the chain is performed. Started with particular examples, it is completely generalized and apparently constitutes by itself a new theoretical result. Apart from that, it has permitted the formulation of the reduced force field for folding in the space of dihedral angles (see Section 3.3), with exclusion of the geometry constraints contained in the classical force fields. Instead, the new model is supplemented by the terms aimed to capture other effects that may be important for successful protein structure predictions.

In order to reproduce the interaction with the solvent, an implicit hydration model was developed in the course of this work. The details of this model are described in Section 3.4. The underlying idea is to estimate the number of water molecules that can directly contact each atom. The suggested strategy is to generate a grid consisting of twelve uniformly distributed points on the surface of each atom. The grid is randomly rotated before each energy evaluation, and the points are checked with regard to their exposure to water. The obtained atomic hydration degrees are used for the evaluation of atomic solvation energies, as well as for the modeling of the electrostatic screening effect.

As described above and more detailed in Chapter 1, findings from molecular biology suggest that protein folding proceeds cotranslationally, starting from a certain initial conformation, which may be important for achievement of the final result. Therefore, the strategy adopted in this work is to generate atomic coordinates for all necessary amino acids, and then to synthesize the desired polypeptide chain in a way maximally resembling the natural one (with presumingly reasonable simplifications). Thus, amino acids are appended one by one, obtaining certain configurations, and each chain elongation step is followed by energy minimization. To prevent the twisting of the nascent polypeptide around the current elongation center, the force field is supplemented by a special term intended to favor folding in a given half-space.

The coordinate transformations necessary for the implementation of the desired process are described together with the related analysis in Section 3.5. It is probably needless to mention that all the transformational analysis, with exception of the definition of rotation about a vector (where an example source was cited), is performed independently in this work, although the basic derivations can be probably found in an equivalent or an alternative form elsewhere, given their applicability in many different areas.

A significant improvement in efficiency can be achieved via analytical calculation of energy derivatives, particularly if their computation during simulations is coupled with energy evaluations. Besides, the visualization of forces significantly facilitates the understanding of the interplay between different interactions. A discussion concerning this issue can be found in Section 3.6. The atomic forces arising from bending and stretching, as well as the energy derivatives with respect to dihedral angles, were derived in course of this work. The subsequent exploration of the literature using corresponding keywords have revealed that some of the described relations can be found in a similar form elsewhere. This fact is not surprising, given the importance of their application. Thus, references [2–4] contain similar expressions. Despite that, the matter of this section is essentially different from the content of the mentioned papers.

Section 3.7 is focused on dynamics of the molecule in dihedral angle space. In this section, equations of motion were derived for constrained dynamics and supplemented by a term that accounts for friction. A brief analysis of the motion equations suggests that a certain scaling for the energy gradient can help to obtain a minimization path resembling a natural folding pathway.

The first section in Chapter 4 describes the developed simulation software. Section 4.2 proceeds with the implementation of the proposed model. It describes the algorithms for determination of the bond weights and for creation of the list of rotatable bonds, as well as for generation of initial atomic coordinates for amino acids. The algorithms rely on the operations for coordinate transformations and the related specific notations, introduced in the preceeding chapter. This section ends with descriptions of the algorithms for chirality correction, discussion of the overall model implementation, and a suggestion for enhancement of computational efficiency. According to the latter, the atoms are grouped with respect to the residue number, and the computations are reduced if the residues are separated by a significant distance.

Simulation results are briefly discussed in Section 4.3. The contributions of different interactions and the discrepancies arising from various permittivity models are elucidated in Subsection 4.3.1. Energy minimization issues are described in Subsection 4.3.2. Simulations have shown that minimization efficiency is significantly improved when minimization is performed in the space of dihedral angles with utilization of the proposed model. By contrast, minimization of the energy given by the force field of the classical type, described in Section 2.2, tends to resolve high-energy configurations largely at the cost of angle bending.

On the other hand, simple minimization that allows only decrease in energy seems to be inappropriate for simulation of protein folding. Therefore, the acceptance criterion for a successful step in the direction of twisting forces was modified to allow certain energy increase. The form of the acceptance criterion was inspired by thermodynamical considerations described in Chapter 1.

The last section is devoted to the concluding remarks and the outlook for further research.

REMARKS ABOUT NOTATIONS

To make reading this text more convenient, certain notational rules were used throughout all chapters. For example, different types of alphabets and font shapes are used to denote matrices $(\mathbf{M})$, vectors $(\vec{v})$ and their components* $(v^{[k]})$, angles[†] (α), sets $(\mathcal{G})$, etc. Upper indices, which refer to vector components or specify some other characteristics[‡], are placed in squire brackets, in order to make them clearly distinguishable from the operation of raising to a certain degree. Lower indices are reserved for ordinal numbering or similar object specification[§]. The usage of upper and lower indexes is demonstrated in the following example:

$$\|\vec{r}_i\| = \sqrt{(r_i^{[1]})^2 + (r_i^{[2]})^2 + (r_i^{[3]})^2}.$$

The notational principles used in the text, even those not mentioned here, should be rather intuitive and easy to figure out.

Specific mathematical notations are explained only when they are introduced for the first time. Additionally, the summary of all such notations is given in Appendix C. General notations, which are assumed to be self-evident, are not defined explicitly in the text. However, they are also listed in Appendix C. An example of such notation is the one used for a scalar product of two vectors.

---

*For the sake of clarity, customary notations x, y, and z were used for Cartesian coordinates in Subsection 1.5.1, related to quantum mechanics.

[†]Additionally, following conventional notations, some other values can be denoted by lowercase Greek letters, such as, for example, electrical permittivity ϵ or electronegativity χ.

[‡]Like in case of angle bending and bond stretching parameters $k_i^{[a]}$ and $k_{ij}^{[b]}$, related to certain atoms A_i and A_j.

[§]For example, $\vec{r}_N$ is to be interpreted as the position of the nitrogen atom in a specific residue determined by the context.

Figures, tables and equations are numbered separately, always independently in each chapter. For definitions, lemmas, propositions, theorems and algorithms the same counter is used, since it significantly facilitates orientation in the text. In any case, a label for an object is constructed from the chapter number together with the internal counter value for this class of objects inside the chapter, separated by a dot. Equation labels are additionally supplemented by parentheses, all other by the object qualifier preceeding the number. An equation is numbered only if it is necessary for a reference.

ACKNOWLEDGMENT

It is probably not very common to mention school teachers in the acknowledgment of a thesis. However, often a school teacher is a one who contributes to a great extent to student's interest towards a certain subject and thereby determines future professional orientation of this person. I got attracted by mathematics and computer science in the middle school, largely during the optional lessons organized by our mathematics teacher for enthusiastic students. Unfortunately, this wonderful teacher passed away two years after I left to another school for more intensive study of mathematics and computer science.

My fascination by life sciences is largely a merit of the Young Academy of Marine Biology (Russian abbreviation – MAMB), which I attended during my last three years in the middle school. This organization was founded by the Institute of Marine Biology of the Far Eastern Branch of the Russian Academy of Sciences. In MAMB, many life science subjects, not limited to marine biology, were taught on the university level to senior pupils in the time free from school lessons. The sincere involvement and competence of our lecturers, which were mostly volunteers, was very motivating for study. In particular, I am very indebted to Lilia G. Kondrashova, Dr. Vadim V. Kumeiko, and Dmitry K. Derzhavin for the knowledge that I still constantly use in my work. At that time I got involved in reading scientific literature, mainly concerning neurobiology, cell biology, and biochemistry. Years passed, but my interest to these areas have never gone out.

This work was inspired by a review paper of Prof. Dr. Arnold Neumaier [5], who outlined the problem of molecular modeling in application to protein structure prediction and gave a critical review of contemporary methods and simulation software related to it. This paper left a large resonance in my mind, being an excellent self-contained introduction to the subject that strongly concerned my long-established interests. I started to work on this problem being a research associate with teaching tasks at the Department of Mathematics of Berlin Technical University. Prof. Dr. Andreas Unterreiter have granted me the freedom to choose my research topic and took care of ordering the necessary literature. I am thankful to the IT-support team there, which promptly fixed all hardware and system-related problems in the course of my work at TU Berlin. I would like to particularly mention Annette Jäkel and Andreas Flügge in this context. Some essential parts of the modeling approach, described in this thesis, and of the related simulation software were developed by me already during that working period. I am very grateful to Lars Oeverdieck for an extension of my contract. At that time and later on, I and my work results have benefited from recommendations of Dr. Werner Benger, including but not limited to those concerning some elegant usage of software utilities.

Inevitably, I am very indebted to Prof. Dr. Dr. h. c. Helmut Neunzert, who helped me many times, first by making possible my study in Germany and latest by giving me an opportunity to come back to Kaiserslautern for writing this thesis. Furthermore, I am immensely thankful to Dr. Falk Triebsch for his valuable help and advice in situations with organizational problems.

I would like to express my deep gratitude to Dr. Patrick Lang, the head of the Department of System Analysis, Prognosis and Control at Fraunhofer Institute for Industrial Mathematics (Fraunhofer ITWM*) in Kaiserslautern, for making possible the completion of this work at ITWM, for the scholarship that allowed me to concentrate solely on this task, before I started to work on other exciting interdisciplinary projects here, and for his guidance during that time.

I am sincerely grateful to Prof. Dr. Tobias Damm for his agreement to supervise my thesis upon my return to Kaiserslautern, although this topic was not among his main interests. I really appreciate his professional and personal qualities, time and patience.

Naturally, I am deeply thankful to my parents, which have always inspired me to engage in science and helped with general suggestions and literature. My supreme thanks are also directed to the rest of my family and friends, which, along with my parents, supported and encouraged me to accomplish this work.

Besides, I would like to take this opportunity to thank those people who organize and make contributions to Wikipedia (`www.wikipedia.org`). Although I did not cite it directly in my thesis, it was always a valuable source of information, which helped me to have a wider overview and a quicker access to referred literature.

---

*In German, Fraunhofer Institut für Techno- und Wirtschaftsmathematik.

CHAPTER 1

Biological, Chemical, and Physical Background

1.1 Introduction

Proteins are essential components of any living cell. They have very diverse functions: catalyze chemical reactions and control gene expression, constitute a cytoskeleton and perform muscle contraction, transport electrons, ions and uncharged molecules, enable recognition of cellular signals or alien invasion. The properties of a protein molecule are determined by its spatial structure (see Fig. 1.1) and location of charged atom groups, which often have to be very specific for a protein to perform a certain function.

The spatial structure of a protein depends on its chemical composition. A protein consists of one or more associated *polypeptide* chains, which are built of consequently

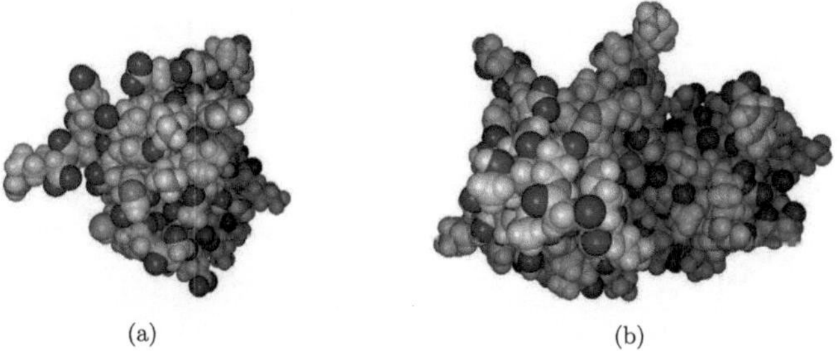

(a)   (b)

FIG. 1.1: Space-filling representations* of two native protein structures determined by nuclear magnetic resonance spectroscopy[†]. (a) Glucocorticoid receptor DNA-binding domain. (b) Bovine pancreatic ribonuclease A.

---

*Molecular images here and further in the text are generated by the program SiViProf, developed in course of this work. For an explanation of the color notations see Table A.1 in Appendix A.

[†]Atomic coordinates are obtained from RCSB Protein Data Bank (see Section 1.10 for details), from records 1GDC by Baumann *et al.* [6] and 2AAS by Santoro *et al.* [7].

connected amino acid residues (see Section 1.2 for details). The number of residues can vary from about fifty to many thousands, depending on protein functions. Shorter sequences are usually referred as *peptides*[*] and often do not have any fixed spatial arrangement in solution.

The sequence of residues is unique for each protein and believed to predetermine the result of folding of the synthesized chain into its native state in a proper environment. The latter statement is named *Anfinsen's dogma* after the Nobel prize laureate Christian B. Anfinsen, who has shown in 1961 that ribonuclease (see Figures 1.1 (b) and 1.2 (a, b)) with reduced disulfide bonds and disrupted tertiary structure was able to restore its enzymatic activity upon removal of the denaturating agent [8]. Later it was proven that also many other small proteins are able to refold *in vitro*[†] into their functional form. These results motivated numerous attempts to compute native three-dimensional structure of proteins based on given amino acid sequences.

Protein structure prediction is a subject of intense research, given the importance of its academic and medical applications. A large number of known protein sequences is already available, and the amount of this data grows rapidly. By contrast, an experimental determination of protein three-dimensional structures by means of X-ray crystallography or nuclear magnetic resonance spectroscopy is relatively expensive and time consuming (see Subsections 1.9.1 and 1.9.2). Computation of native protein structures from their amino acid sequences could contribute to the understanding of the organization of living organisms on the molecular level and give a clue to treatment of many diseases. It would also enable more rational drug design, helping to lower the costs and the amount of time required to introduce new medications.

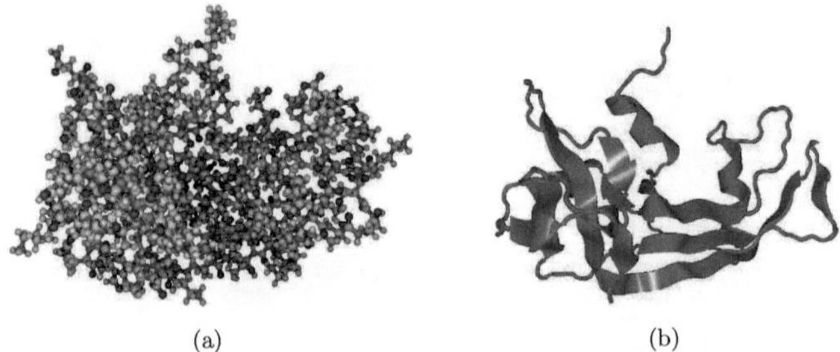

(a)   (b)

FIG. 1.2: Some more insight into the structure of bovine pancreatic ribonuclease A[‡]: (a) the ball-and-stick model visualizing all atoms and bonds between them, (b) the ribbon model showing the fold of the main chain. A discussion about different models for protein visualization follows further in the text, see Subsections 1.5.4 and 1.6.2.

---

[*]Peptides consisting of two, three, or a few amino acid residues are called *dipeptides*, *tripeptides* or *oligopeptides* respectively.

[†]Outside a living organism, literally, *in glass* (Latin).

[‡]Atomic coordinates are the same as in Figure 1.1 (b).

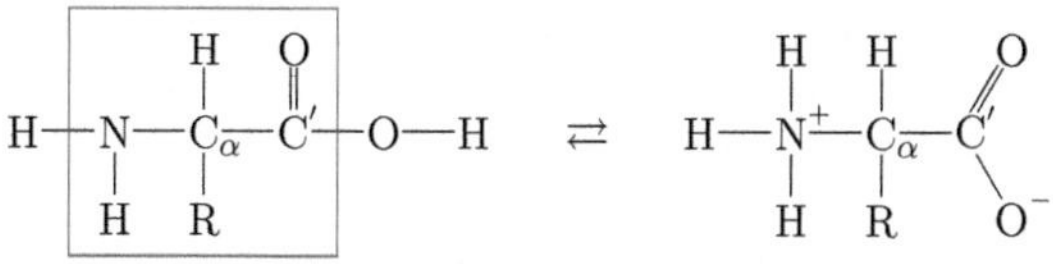

FIG. 1.3: General structure of amino acids in a nonionized and an ionized form. R stands for a side chain. The red box outlines a non-terminal residue after inclusion into a polypeptide.

1.2 CHEMICAL STRUCTURE OF PROTEINS

There are 20 different amino acids that are used by cells as building blocks in protein synthesis. They are listed in Tables 1.1-1.3. All of them, with exception of proline, have a common part, containing an *amino* ($-NH_2$) and a *carboxyl* ($-COOH$) group (Fig. 1.3). The carbon atom of the carboxyl group is conventionally marked as C' in order to distinguish it from the α-carbon bonded to the amino group. The distinctive part of an amino acid is called the *side chain*. Its non-hydrogen atoms are labeled using subsequent Greek letters, starting from the atom linked to C_α. In case of branching, letters are additionally supplied by indexes (see Table A.2 in Appendix A for details).

In course of protein synthesis the common fragments of amino acids are joined together by *peptide bonds*, thereby constituting the *main chain*, or the *protein backbone* (Fig. 1.4). As a result of peptide bond formation, the hydroxyl group ($-OH$) at C' and a hydrogen from the amino group of the next amino acid are removed. Although the structure of proline is somewhat different, it allows its molecules to be incorporated into a chain in a similar way (see Subsection 1.6.1).

The residues are appended to the carboxyl end in a certain order, prescribed by the corresponding genetic code. This procedure is termed *translation*. The details of the protein synthesis that are relevant for initial arrangement of atoms in a nascent protein are discussed in Section 1.8. After completion of translation, some chemical alternations of standard amino acids can be performed as a part of a controlled process, termed *posttranslational modification*.

The sequence of residues in a protein is called its *primary structure*. It is written using conventional one- or three-letter abbreviations (see Table A.2 in Appendix A), starting from the amino end. The reverse order of residues corresponds to another protein.

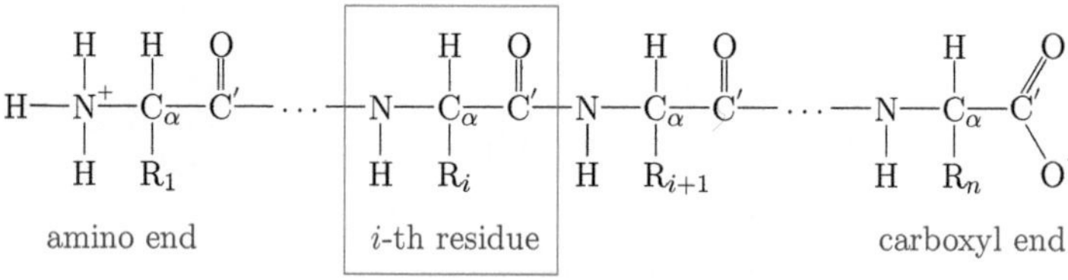

FIG. 1.4: A polypeptide chain. Peptide bonds connect C' and N atoms.

Once or even before the polypeptide chain is completely synthesized, it adopts a certain conformation, which is responsible for the protein functions. The folding pathway and the resulting structure are largely determined by the properties of the constituent amino acid residues. For example, if folding happens in cytosol, which represents mainly a mixture of water with salts, nonpolar side chains seek to avoid

TABLE 1.1: Amino acids with hydrophilic ionizable side chains.

Name	Chemical formula*	Molecular structure[†]
Aspartic acid	$^+H_3N\!-\!\underset{\underset{H}{\mid}}{\overset{\overset{COO^-}{\mid}}{C}}\!-\!CH_2\!-\!COO^-$	
Glutamic acid	$^+H_3N\!-\!\underset{\underset{H}{\mid}}{\overset{\overset{COO^-}{\mid}}{C}}\!-\!CH_2\!-\!CH_2\!-\!COO^-$	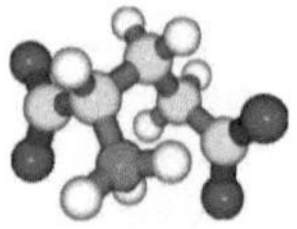
Cysteine	$^+H_3N\!-\!\underset{\underset{H}{\mid}}{\overset{\overset{COO^-}{\mid}}{C}}\!-\!CH_2\!-\!SH$	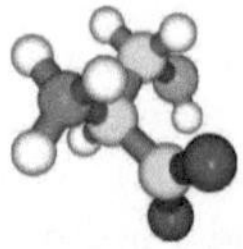
Tyrosine	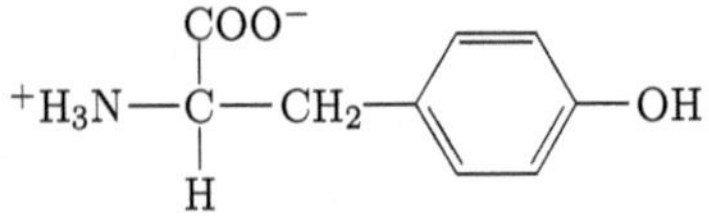 $^+H_3N\!-\!\underset{\underset{H}{\mid}}{\overset{\overset{COO^-}{\mid}}{C}}\!-\!CH_2\!-\!\langle\!\bigcirc\!\rangle\!-\!OH$	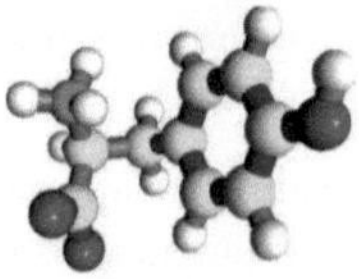
Histidine	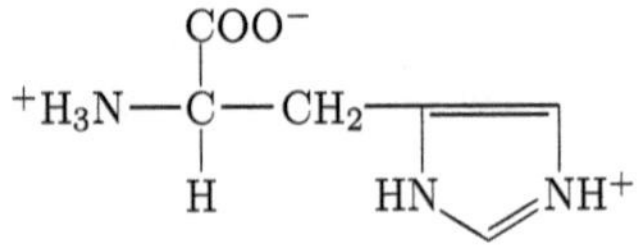 $^+H_3N\!-\!\underset{\underset{H}{\mid}}{\overset{\overset{COO^-}{\mid}}{C}}\!-\!CH_2\!-$ imidazole (HN, NH^+)	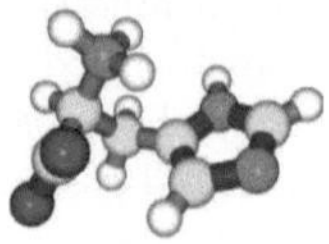
Lysine	$^+H_3N\!-\!\underset{\underset{H}{\mid}}{\overset{\overset{COO^-}{\mid}}{C}}\!-\!(CH_2)_4\!-\!NH_3^+$	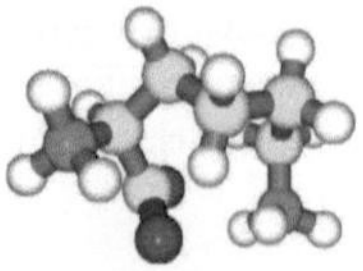
Arginine	$^+H_3N\!-\!\underset{\underset{H}{\mid}}{\overset{\overset{COO^-}{\mid}}{C}}\!-\!(CH_2)_3\!-\!NH\!-\!\underset{\underset{NH_2}{\mid}}{C}\!\!=\!\!NH_2^+$	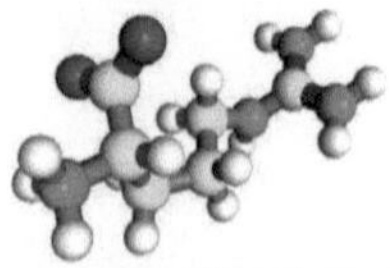

*The presented form considerably prevails in physiological conditions, except for histidine: only about one fourth of histidine side chains is protonated in cytosol (see Section 1.4).

[†]In the last column the histidine side chain is depicted in the more probable non-protonated form. Single and double bonds in conjugated systems are treated as bonds having partial double character.

TABLE 1.2: Amino acids with hydrophobic side chains.

Name	Chemical formula	Molecular structure
Proline	^+H_2N — ... COO^-	
Glycine	^+H_3N — $C(H)(H)$ — COO^-	
Alanine	^+H_3N — $C(H)(CH_3)$ — COO^-	
Valine	^+H_3N — $C(H)$ — COO^-, CH—CH_3, CH_3	
Leucine	^+H_3N — $C(H)$ — COO^-, CH_2—CH—CH_3, CH_3	
Isoleucine	^+H_3N — $C(H)$ — COO^-, CH—CH_2—CH_3, CH_3	
Methionine	^+H_3N — $C(H)$ — COO^-, CH_2—CH_2—S—CH_3	
Phenylalanine	^+H_3N — $C(H)$ — COO^-, CH_2—C_6H_5	
Tryptophan	^+H_3N — $C(H)$ — COO^-, CH_2—(indole)	

TABLE 1.3: Amino acids with hydrophilic non-ionizable side chains.

Name	Chemical formula	Molecular structure
Serine	$^+H_3N-\underset{\underset{H}{\mid}}{\overset{\overset{COO^-}{\mid}}{C}}-CH_2-OH$	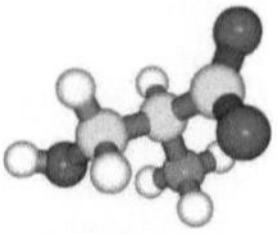
Threonine	$^+H_3N-\underset{\underset{H}{\mid}}{\overset{\overset{COO^-}{\mid}}{C}}-\underset{\underset{CH_3}{\mid}}{CH}-OH$	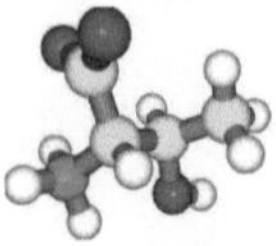
Asparagine	$^+H_3N-\underset{\underset{H}{\mid}}{\overset{\overset{COO^-}{\mid}}{C}}-CH_2-C\overset{O}{\underset{NH_2}{}}$	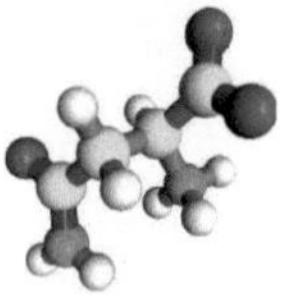
Glutamine	$^+H_3N-\underset{\underset{H}{\mid}}{\overset{\overset{COO^-}{\mid}}{C}}-CH_2-CH_2-C\overset{O}{\underset{NH_2}{}}$	

contact with surrounding solvent. This causes a collapse of a nascent polypeptide into a molten globule with its subsequent compactization. Polar groups stabilize the conformation by building hydrogen and disulfide bonds. They are also responsible for specific binding to other molecules. On the other hand, in absence of hydrophilic residues, which tend to stay in contact with water, an adhesion to neighboring protein molecules would be inevitable, much like oil collects in drops being mixed with water.

The division of the side chains into groups of hydrophobic and hydrophilic is not strict. Some residues have intermediate properties. For example, tyrosine is sometimes classified as hydrophobic, since its aromatic ring favors aggregation with non-polar molecules. The side chains of glycine and alanine are relatively small and therefore fit into both hydrophobic and hydrophilic environment.

1.3 CHIRALITY OF AMINO ACIDS

All standard amino acids, with exception of glycine, can occur in form of at least two optical* isomers, which differ like left and right hands (Fig. 1.5). In the L-form, the nitrogen, the carboxyl carbon, and the side chain appear in the clockwise order,

*The term *optical activity* refers to the property of chiral compounds to rotate the plane of polarized light.

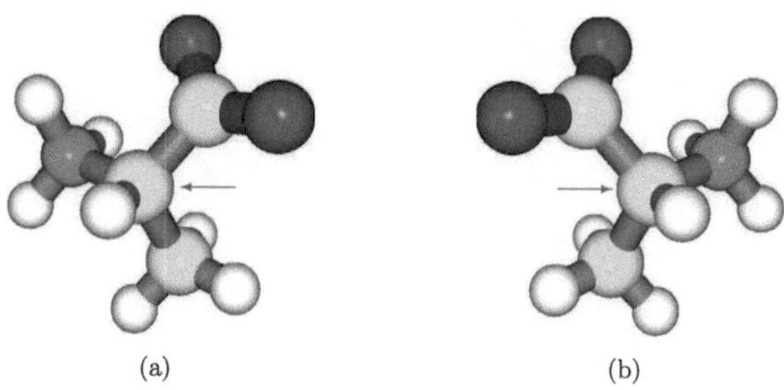

FIG. 1.5: L-alanine (a) and D-alanine (b). The chiral center at the α-carbon is pointed by a red arrow.

when viewed along the bond from the hydrogen to α-carbon. In D-isomers they show up in the counterclockwise order. For glycine these two forms are equivalent, since its side chain consists of only one hydrogen atom.

Proteins are composed of only L-amino acids. D-alanine and D-glutamic acid are usual components of peptidoglican cell walls of bacteria, blue-green algae and some viruses. The presence of D-isomers makes a shell resistant to decomposition by most peptidases. D-residues are also found in short peptides isolated from amphibian skin, snail ganlia and venom of a number of exotic species [9–11]. These inclusions, however, are not translated in the usual way, but either integrated into the chain by certain enzymes or derived posttranslationally from corresponding L-isomers [9, 12].

Transitions between L- and D-forms almost do not occur spontaneously: the rate of such conversion is so slow that it can be used for dating of biological fossil objects, in combination with other techniques.

Any atom with four different *substitutes*, i.e. groups attached to it, may be an origin of optical activity. It is called *chiral center* and often referred by a more broad term *stereocenter*. Apart from the described center of optical activity at C_α, molecules of isoleucine and threonine have another such center at β-carbon. The side chains of these amino acids can occur in the S- or R-form, named following another, more common, nomenclature, which is based on the Cahn Ingold Prelog priority rules. According to this nomenclature, certain priorities are assigned to the atom groups, bound to chiral center. First, the atoms directly attached to the stereocenter are compared. Atoms with higher atomic numbers obtain higher priorities. For the groups, having the same element bound to the chiral center, further atoms in the chains are considered, until different atoms are found. When the priorities are assigned, the S- or R-form are distinguished by the order of substitutes, viewed from the group with the lowest priority to the chiral center. In the S-form, groups appear in the clockwise order from the higher to the lower priority.

Normally only the L-isoleucine and L-threonine with the side chains in the S- and R-configuration respectively (Fig. 1.6(a, d)) participate in protein synthesis.

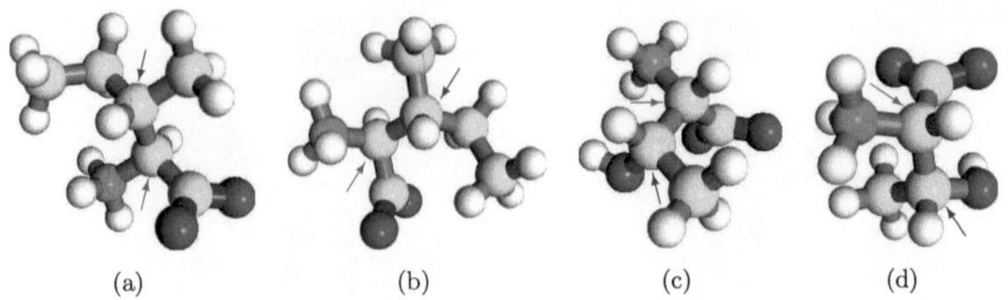

(a)   (b)   (c)   (d)

FIG. 1.6: Amino acids with two chiral centers. L-isoleucine with the side chain in the S- (a) and R-configuration (b). L-threonine with the side chain in the S- (c) and R-configuration (d). The chiral centers at the α-carbons and the chiral centers of the side chains are pointed by red and blue arrows respectively.

1.4   IONIZATION OF AMINO ACID RESIDUES

To estimate the probability of ionizable atom groups to be in one or the other state at physiological conditions, we first consider dissociation of a simple monoprotic acid HA:

$$\mathrm{HA} \underset{k_2}{\overset{k_1}{\rightleftharpoons}} \mathrm{H}^+ + \mathrm{A}^-, \tag{1.1}$$

where k_1 and k_2 denote specific reaction rates. According to the law of mass action, first formulated by Waage and Guldberg in 1864 [13], the reaction rate is proportional to the product of the reactant masses raised in degrees of their stoichiometric coefficients*. In case of a non-dilute solution one has to take into consideration ionic activities, but we can think of an ideal solution.

Hence the change of the anion concentration can be described by the differential equation

$$\frac{d\,[\mathrm{A}^-]}{dt} = k_1[\mathrm{HA}] - k_2[\mathrm{H}^+][\mathrm{A}^-]. \tag{1.2}$$

In equilibrium there is no concentration change, therefore we set $\frac{d\,[\mathrm{A}^-]}{dt} = 0$ in equation (1.2) and obtain:

$$\frac{[\mathrm{H}^+][\mathrm{A}^-]}{[\mathrm{HA}]} = \frac{k_1}{k_2} =: K_a. \tag{1.3}$$

The stated in equation (1.3) consequence of the mass action law is often termed as

---

*For example, in a reaction involving n molecules of A and m molecules of B,

$$\mathrm{nA} + \mathrm{mB} \longrightarrow \mathrm{A_n B_m}\,,$$

the rate of product formation is

$$\frac{d\,[\mathrm{A_n B_m}]}{dt} = k[\mathrm{A}]^n[\mathrm{B}]^m$$

with a positive constant k. The members in square brackets denote concentrations of corresponding substances.

the law of mass action itself. *Acid dissociation constant** K_a can be also utilized to characterize the protonation of a base B through the dissociation of the conjugate acid BH^+:

$$K_a = \frac{[H^+][B]}{[BH^+]}.$$

Often $pK_a := -\lg K_a$ is used for convenience instead of K_a, as $pH = -\lg[H^+]$ is adopted for measurement of the solution acidity[†]. pK_a, as well as pH, depend on temperature. One can see that pK_a is equal to pH at which the concentrations of the protonated and non-protonated form of the substance are equal, i.e. the considered atom group is ionized in 50% of molecules.

Therefore, the probabilities of the charged and uncharged state relate for an acidic group as

$$\frac{[A^-]}{[HA]} = \frac{K_a}{[H^+]} = 10^{\,pH-pK_a} \tag{1.4}$$

and for a basic group as

$$\frac{[BH^+]}{[B]} = \frac{[H^+]}{K_a} = 10^{\,pK_a-pH}. \tag{1.5}$$

It is natural to assume that folding occurs at a constant pressure, and that the system exchanges heat with its environment. We shall recall that a steady state under isotherm-isobaric conditions is achieved when the Gibbs energy is minimized. The *Gibbs free energy* G is a thermodynamic potential given by

$$G := E + pV - TS, \tag{1.6}$$

also written as

$$G := H - TS. \tag{1.7}$$

Here E denotes the internal energy of the system, p – the pressure, V – the volume, T – the absolute temperature, and S – the entropy. $H := E + pV$ is the enthalpy.

The *entropy* S is a measure of randomness of the molecular ordering in the system. Different definitions of entropy are possible (see, for example, [1, 14, 15]). For our purposes we shall define it as follows: the entropy of a certain state of the system is given by

$$S = k_B \ln W, \tag{1.8}$$

where k_B is the Boltzmann's constant, and W is the number of ways in which the given configuration of the system can be achieved. In application to protein folding, one usually talks not about a single particle or molecule, but about a mole of such objects, and the related energies are measured in kcal/mol. The corresponding molar entropy is given by

$$S = \bar{R} \ln W,$$

---

[*]For non-dilute solutions K_a defined by equation (1.3) is concentration-dependent and sometimes called *apparent acid dissociation constant*.

[†]pH = 7.0 in clean water at 25° C. Acidic solutions have smaller pH values, basic – larger values.

where

$$\bar{R} := k_B N_A$$

is the gas constant, and N_A is the Avogadro's number.

A related notion is the *Helmholtz free energy*, which is defined as

$$F := E - TS.$$

Since the value pV, and thus the difference between G and F, is negligible for the considered systems [1], we shall often use the term *free energy* without further specifications.

To evaluate the contribution of the ionization to the molar Gibbs free energy of a molecule, let us now look at the free energy of the mixture of ionized and nonionized groups of the same type:

$$G = g_i \nu_i + g_n \nu_n - T S_{\text{mix}}.$$

Here g_i and g_n denote the molar free energy of an ionized and a nonionized group respectively, ν_i and ν_n are the molar amounts of groups in the corresponding forms, T is the absolute temperature and S_{mix} is the entropy of mixing.

The latter can be computed by substituting for W in (1.8) the number of different ways to select $\nu_i N_A$ groups among $(\nu_i + \nu_n) N_A$ groups for ionization. Thus,

$$S_{\text{mix}} = k_B \ln \left(\frac{((\nu_i + \nu_n) N_A)!}{(\nu_i N_A)!(\nu_n N_A)!} \right).$$

Using the Stirling's approximation,

$$\ln n! \approx n \ln n - n + \frac{1}{2} \ln(2\pi n), \quad n > 10, \tag{1.9}$$

we obtain:

$$S_{\text{mix}} \approx k_B \left(N_A(\nu_i + \nu_n) \ln(N_A(\nu_i + \nu_n)) - N_A(\nu_i + \nu_n) + \frac{1}{2} \ln(2\pi N_A(\nu_i + \nu_n)) - \right.$$

$$- N_A \nu_i \ln(N_A \nu_i) + N_A \nu_i - \frac{1}{2} \ln(2\pi N_A \nu_i) -$$

$$\left. - N_A \nu_n \ln(N_A \nu_n) + N_A \nu_n - \frac{1}{2} \ln(2\pi N_A \nu_n) \right) =$$

$$= \bar{R} \left(\nu_i \ln \left(\frac{\nu_i + \nu_n}{\nu_i} \right) + \nu_n \ln \left(\frac{\nu_i + \nu_n}{\nu_n} \right) \right).$$

In fact, the error introduced by the Stirling's approximation (1.9) can become arbitrary small with growing n (see, for example, [16], p. 511). We assume that the system includes a sufficiently large number of groups from each category, so that the approximation error can be neglected. Hence follows that

$$G = g_i \nu_i + g_n \nu_n + \bar{R} T \left(\nu_i \ln \nu_i + \nu_n \ln \nu_n - (\nu_i + \nu_n) \ln(\nu_i + \nu_n) \right).$$

The free energy change upon ionization of $\Delta\nu$ mole of nonionized groups is

$$\Delta G = g_i \Delta\nu - g_n \Delta\nu + \bar{R}T \left((\nu_i + \Delta\nu) \ln(\nu_i + \Delta\nu) - \right.$$
$$\left. -\nu_i \ln\nu_i + (\nu_n - \Delta\nu) \ln(\nu_n - \Delta\nu) - \nu_n \ln\nu_n \right).$$

For both positive and negative $\Delta\nu$ $(\Delta\nu << \nu_f)$ holds:

$$\lim_{\Delta\nu\to 0} \frac{(\nu_f + \Delta\nu) \ln(\nu_f + \Delta\nu) - \nu_f \ln\nu_f}{\Delta\nu} = \frac{d(\nu_f \ln\nu_f)}{d\nu_f} = \ln\nu_f + 1,$$

where f stands for both the ionized and nonionized form. Therefore follows:

$$\lim_{\Delta\nu\to 0} \frac{\Delta G}{\Delta\nu} = g_i - g_n + \bar{R}T \ln\left(\frac{\nu_i}{\nu_n}\right). \tag{1.10}$$

In equilibrium, the right hand side of equation (1.10) is zero, hence

$$g_i - g_n + \bar{R}T \ln\left(\frac{\nu_i}{\nu_n}\right) = 0.$$

Using equations (1.4) and (1.5) we obtain the following expressions for evaluation of the molar free energy of ionization:

$$g_i - g_n = -\bar{R}T \ln \frac{K_a}{[\mathrm{H}^+]} = \bar{R}T \ln 10 \, (\mathrm{pH} - \mathrm{p}K_a) \tag{1.11}$$

for an acidic group and

$$g_i - g_n = -\bar{R}T \ln \frac{[\mathrm{H}^+]}{K_a} = \bar{R}T \ln 10 \, (\mathrm{p}K_a - \mathrm{pH}) \tag{1.12}$$

for a basic group.

Ionizable groups that can be found in proteins are listed in Table 1.4 together with intervals for their $\mathrm{p}K_a$ values, estimated probabilities to be ionized in normal conditions, and the corresponding ionization free energies. Apparently, the most groups preferably stay in one particular state, which is shown in Table 1.2. Histidine, cysteine, and the terminal amino group have $\mathrm{p}K_a$ values close to the pH of cytosol, therefore appear in adequate amounts in the both forms. Histidine is often used in active centers of enzymes due to its sensitivity to local environmental changes [1]. A couple of cysteine residues can build a *disulfide bridge* (-S-S-), which stabilizes the structure of the protein molecule. Ionization of the thiol groups (-SH) of cysteine is an important step in disulfide bond formation and exchange. However, disulfide bridges are usually unstable in cytosol and mostly occur in secretory proteins.

One should bear in mind that the ionization behavior is affected by electric permittivity and ionic strength of the ambience. During the folding process the ionizable groups can get surrounded by nonpolar side chains or come into a neighborhood of a number of charged groups. As a result, actual values can in some cases significantly deviate from the ones given in the table.

TABLE 1.4: The pK_a values of ionizable groups at 25° C, the corresponding probabilities $P_{\pm}$ of the charged states at neutral pH, and the free energy change ΔG_i upon ionization.

Atom group[*]	Residue	pK_a[†]	$P_{\pm}$[‡]	ΔG_i[§] (kcal/mol)
$-C'OOH \leftrightarrow -C'OO^-$	carboxyl end	$[3.4, 3.8]$	$(0.9993, 0.9998)$	$(-5.0, -4.3)$
$-C_\gamma OOH \leftrightarrow -C_\gamma OO^-$	aspartic acid	$[3.9, 4.0]$	$(0.9990, 0.9993)$	$(-4.3, -4.0)$
$-C_\delta OOH \leftrightarrow -C_\delta OO^-$	glutamic acid	$[4.4, 4.5]$	$(0.9968, 0.9975)$	$(-3.6, -3.4)$
$\geqslant N \leftrightarrow \geqslant NH^+$	histidine	$[6.3, 6.6]$	$(0.1663, 0.2848)$	$(0.5, 1.0)$
$-NH_2 \leftrightarrow -NH_3^+$	amino end	$[7.4, 7.5]$	$(0.7152, 0.7598)$	$(-0.7, -0.5)$
$-SH \leftrightarrow -S^-$	cysteine	$[7.5, 9.5]$	$(0.0031, 0.2403)$	$(0.6, 3.5)$
$-O_\zeta H \leftrightarrow -O_\zeta^-$	tyrosine	$[9.6, 10.0]$	$(0.0009, 0.0026)$	$(3.5, 4.1)$
$-N_\zeta H_2 \leftrightarrow -N_\zeta H_3^+$	lysine	$[10.0, 10.4]$	$(0.9990, 0.9997)$	$(-4.7, -4.0)$
$=NH \leftrightarrow =NH_2^+$	arginine	> 12.5	> 0.999996	< -7.4

1.5   INTRAMOLECULAR INTERACTIONS

To build a reasonable model for protein structure prediction, it is important to consider the types of atomic interactions that are encountered in proteins and may have a substantial influence on protein structure. We shall start our discussion with covalent bonds, which are typically the strongest interaction type, then consider electrostatic interactions between atoms that are not bound to each other. After that we shall discuss van der Waals interactions, which prevent excessive convergence of non-bound atoms, even with opposite charges, but result in a relatively weak attractive force when the atoms are separated by a sufficient distance. Finally, we discuss less abundant hydrogen bonds, which represent another type of non-covalent interactions playing an important role in stabilization of a protein structure.

The properties of chemical elements, among which are the number of formed bonds and the values of bond angles, are determined by features of atomic electron shells. These and some other properties can be explained by means of quantum mechanics.

---

[*]The Greek letters are used to eliminate the ambiguity in the atom group indication. The details of the atom numeration can be found in the Appendix A.

[†]The pK_a values are taken from [17]

[‡]The probabilities $P_{\pm}$ are calculated according to the formula

$$P_{\pm} = \frac{P_{\pm/0}}{1 + P_{\pm/0}},$$

where $P_{\pm/0}$ is the probability ratio computed from equation (1.4) or (1.5) for an acidic and a basic group respectively.

[§]The molar free energy changes upon ionization, $\Delta G_i := g_i - g_n$, are evaluated using equations (1.11) and (1.12).

1.5.1 PRINCIPLES OF QUANTUM MECHANICS AND ATOMIC ORBITALS

The movement of a particle of the mass m in a time-independent force field is described by the *time-independent Schrödinger equation*:

$$-\frac{\hbar^2}{2m}\Delta\Psi + U\Psi = E\Psi. \tag{1.13}$$

Here $\Psi : \mathbb{R}^3 \to \mathbb{C}$ is the *wavefunction* of the particle with a constant total energy $E \in \mathbb{R}$, $U : \mathbb{R}^3 \to \mathbb{R}$ is the potential energy of the particle, and $\hbar$ is the reduced Planck's constant.

According to the Born interpretation of the wavefunction, if Ψ is normalized such that the integral of $\overline{\Psi}\Psi$ over all space is equal to one, the product $\overline{\Psi}\Psi = |\Psi|^2$ represents the probability density of the particle to be found in a certain position. Thus, physically acceptable wavefunctions must not be equal to zero everywhere, must be square-integrable, continuous and have a continuous slope.

The general form of the Schrödinger equation describes the evolution of a system that includes N particles:

$$\hat{H}\Psi = i\hbar\frac{\partial\Psi}{\partial t}, \tag{1.14}$$

where $\hat{H}$ is the *Hamiltonian operator* of the system, and Ψ is now a function of $3N$ space coordinates and time.

For such a system, the expression

$$|\Psi(x_1, y_1, z_1, \ldots, x_N, y_N, z_N, t)|^2 dx_1\, dy_1\, dz_1 \ldots dx_N\, dy_N\, dz_N \tag{1.15}$$

gives the probability at time t to find simultaneously the first particle in the infinitesimal volume at the point $(x_1, y_1, z_1)^{\mathrm{T}}$, the second one at $(x_2, y_2, z_2)^{\mathrm{T}}$, and so on for the other particles*. In this case, the normalization condition for Ψ takes the form:

$$\int_{-\infty}^{\infty}\int_{-\infty}^{\infty}\int_{-\infty}^{\infty}\ldots\int_{-\infty}^{\infty}\int_{-\infty}^{\infty}\int_{-\infty}^{\infty} |\Psi|^2 dx_1\, dy_1\, dz_1 \ldots dx_N\, dy_N\, dz_N = 1$$

We are interested in a steady state solution, therefore equation (1.14) reduces to the following:

$$\hat{H}\Psi = E\Psi. \tag{1.16}$$

For a system consisting of one particle,

$$\hat{H} = -\frac{\hbar^2}{2m}\Delta + U, \tag{1.17}$$

in accordance with equation (1.13).

---

*Here the positions in Cartesian coordinates are implied.

One can readily see that (1.16) is an eigenvalue equation, and the wavefunctions are the eigenfunctions of the total energy E. The latter can take only discrete values, which are eigenvalues of (1.16) under the restrictions on acceptable solutions along with particular boundary conditions. In other words, the energy is *quantized*, and different wavefunctions correspond to certain energy levels[*].

The Hamiltonian operator $\hat{H}$ was named so by analogy with the Hamiltonian function H from classical mechanics. We shall recall that for a system consisting of a single particle,

$$H(\vec{\mathbf{p}}, \vec{\mathbf{r}}) = \frac{1}{2m}\, \vec{\mathbf{p}} \cdot \vec{\mathbf{p}} + U(\vec{\mathbf{r}}), \tag{1.18}$$

where $\vec{\mathbf{p}}$ is the linear momentum, $\vec{\mathbf{r}}$ is the position, and U is the potential energy of the particle. The first term on the right-hand side of equation (1.18) is the kinetic energy, and respectively the Hamiltonian function gives the total energy.

Similarly, the Hamiltonian operator $\hat{H}$ corresponds to the total energy E. Moreover, according to a fundamental postulate of quantum mechanics, for every physical property in classical mechanics there is a quantum-mechanical operator. In particular, the operators

$$\hat{p}_x = \frac{\hbar}{i}\frac{\partial}{\partial x},$$
$$\hat{p}_y = \frac{\hbar}{i}\frac{\partial}{\partial y},$$
$$\hat{p}_z = \frac{\hbar}{i}\frac{\partial}{\partial z}$$

correspond to the Cartesian components of the particle's linear momentum, and

$$\hat{E}_K = \frac{1}{2m}\left(\hat{p}_x^2 + \hat{p}_y^2 + \hat{p}_z^2\right) =$$
$$= \frac{1}{2m}\left(\left(\frac{\hbar}{i}\right)^2 \frac{\partial^2}{\partial x^2} + \left(\frac{\hbar}{i}\right)^2 \frac{\partial^2}{\partial y^2} + \left(\frac{\hbar}{i}\right)^2 \frac{\partial^2}{\partial z^2}\right) =$$
$$= -\frac{\hbar^2}{2m}\Delta$$

is the operator corresponding to the kinetic energy of the particle, in a striking analogy with classical mechanics.

A hydrogen atom is a system consisting of a nucleus with the positive charge of $1\,|\mathrm{e}^-|$ and an electron moving around it. In fact, the nucleus moves also, although much more slowly than the electron. The wavefunction of the system depends now on the positions of the electron, $(x_e, y_e, z_e)^{\mathrm{T}}$, and nucleus, $(x_n, y_n, z_n)^{\mathrm{T}}$. The Hamiltonian operator for the described system is:

$$\hat{H} = -\frac{\hbar^2}{2m_e}\left(\frac{\partial^2}{\partial x_e^2} + \frac{\partial^2}{\partial y_e^2} + \frac{\partial^2}{\partial z_e^2}\right) - \frac{\hbar^2}{2m_n}\left(\frac{\partial^2}{\partial x_n^2} + \frac{\partial^2}{\partial y_n^2} + \frac{\partial^2}{\partial z_n^2}\right) - \frac{q_e^2}{4\pi\epsilon_0 r}, \tag{1.19}$$

---

[*]Although, certain conditions may allow a continuous energy spectrum.

where m_e and m_n are respectively the masses of the electron and nucleus, q_e is the elementary charge*, ϵ_0 is the vacuum permittivity, and r is the distance between the two particles. The first and second term at the right-hand side of equation (1.19) are the operators for the kinetic energies of the electron and nucleus, and the last term represents the potential energy of the Coulomb interaction between them.

As discussed below, the Schrödinger equation for this system can be solved analytically. For many-electron atoms additional complications arise due to interaction between electrons, and in such cases one is forced to search for approximate solutions. A certain insight into the features of electron shells can be gained if one ignores interelectronic interactions and as an initial approximation considers a model atom with a single electron moving about a nucleus with the charge Zq_e. For the latter problem setup, the potential energy term in (1.19) takes the form:

$$U(r) = -\frac{Zq_e^2}{4\pi\epsilon_0 r}. \tag{1.20}$$

One can show that the Schrödinger equation for the given system of two particles can be subdivided into two parts, one of which can be roughly interpreted as describing the movement of the electron relative to the nucleus, and the other one gives the translational motion of the whole atom in space [14]. In a more precise formulation, each of the two equations obtained after separation concerns one of two fictitious particles, first of which has the mass

$$\underline{m} = \frac{m_e m_n}{m_e + m_n} \tag{1.21}$$

and moves in the central force field with the potential energy given by (1.20), while the other one with the mass

$$\overline{m} = m_e + m_n \tag{1.22}$$

is not subjected to any forces (see, for example, [18] for more details).

Taking $m_e \approx 5.4858 \times 10^{-4}$ Da, for a hydrogen atom with the nucleus mass $m_n \approx 1.0073$ Da we obtain $\underline{m} \approx 5.4828 \times 10^{-4}$ Da, which is about 99.95 % of the electron mass. For heavier nuclei the so-called *reduced mass* $\underline{m}$ approaches m_e even closer, thereby fortifying the basis for the above given interpretation.

Thus, to explore the movement of the electron relative to the nucleus, we shall solve equation (1.13) with $m = \underline{m}$ and U given by (1.20).

Due to the spherical symmetry of the problem, it is natural to use spherical coordinates (r, φ, ϑ), which are conventionally defined as shown in Figure 1.7.

---

*Despite the fact that for quantum mechanical computations it is convenient to introduce a unit system, in which q_e becomes a unit charge (i.e., using $1\,|e^-|$ as a charge unit),we have to use a unit-independent notation for the elementary charge to keep track of correct dimensions in mathematical expressions.

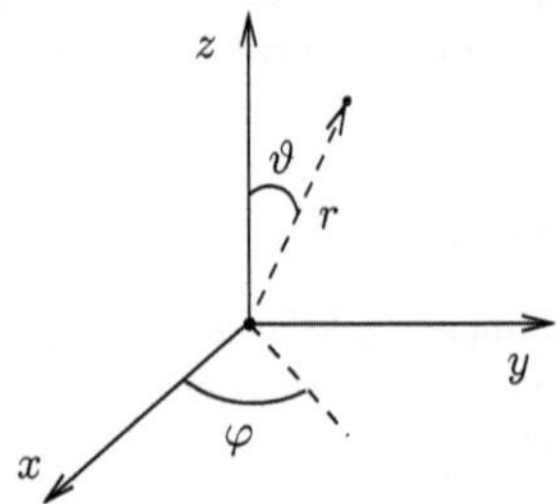

FIG. 1.7: Relation of the spherical coordinates (r, φ, ϑ) to the Cartesian (x, y, z).

To be more specific, we set:

$$r = \sqrt{x^2 + y^2 + z^2}, \tag{1.23}$$

$$\varphi = \begin{cases} 0, & \text{if } x = 0,\ y = 0, \\ \frac{\pi}{2}, & \text{if } x = 0,\ y > 0, \\ -\frac{\pi}{2}, & \text{if } x = 0,\ y < 0, \\ \arctan \frac{y}{x}, & \text{if } x > 0, \\ \pi + \arctan \frac{y}{x}, & \text{if } x < 0. \end{cases} \tag{1.24}$$

$$\vartheta = \begin{cases} 0, & \text{if } x^2 + y^2 + z^2 = 0, \\ \arccos \frac{z}{\sqrt{x^2+y^2+z^2}} & \text{otherwise.} \end{cases} \tag{1.25}$$

For simplicity we shall keep the same notation for the wavefunctions expressed in spherical coordinates, $\Psi = \Psi(r, \varphi, \vartheta)$.

The Laplace-Operator in spherical coordinates (1.23)-(1.25) is given by:

$$\Delta \Psi = \frac{\partial^2 \Psi}{\partial r^2} + \frac{2}{r} \frac{\partial \Psi}{\partial r} + \frac{1}{r^2} \frac{\partial^2 \Psi}{\partial \vartheta^2} + \frac{\cos \vartheta}{r^2 \sin \vartheta} \frac{\partial \Psi}{\partial \vartheta} + \frac{1}{r^2 \sin^2 \vartheta} \frac{\partial^2 \Psi}{\partial \varphi^2},$$

therefore we obtain:

$$\frac{\partial^2 \Psi}{\partial r^2} + \frac{2}{r} \frac{\partial \Psi}{\partial r} + \frac{1}{r^2} \frac{\partial^2 \Psi}{\partial \vartheta^2} + \frac{\cos \vartheta}{r^2 \sin \vartheta} \frac{\partial \Psi}{\partial \vartheta} + \frac{1}{r^2 \sin^2 \vartheta} \frac{\partial^2 \Psi}{\partial \varphi^2} + \frac{2m}{\hbar^2} \left(E + \frac{Z q_e^2}{4\pi\epsilon_0 r} \right) \Psi = 0.$$

With the ansatz

$$\Psi(r, \varphi, \vartheta) = R(r) \Phi(\varphi) \Theta(\vartheta)$$

and subsequent separation of variables we arrive to the equations:

$$r^2 \frac{R''(r)}{R(r)} + 2r \frac{R'(r)}{R(r)} + \frac{2mr}{\hbar^2} \left(Er + \frac{Z q_e^2}{4\pi\epsilon_0} \right) = \lambda_1, \tag{1.26}$$

$$\sin^2 \vartheta \frac{\Theta''(\vartheta)}{\Theta(\vartheta)} + \sin \vartheta \cos \vartheta \frac{\Theta'(\vartheta)}{\Theta(\vartheta)} + \lambda_1 \sin^2 \vartheta = \lambda_2, \tag{1.27}$$

$$\Phi''(\varphi) + \lambda_2 \Phi(\varphi) = 0, \tag{1.28}$$

where λ_1, $\lambda_2 \in \mathbb{C}$ are certain constants, which appear due to separation of variables.

Equation (1.28) is readily solved with the standard ansatz for linear ordinary differential equations with constant coefficients:

$$\Phi(\varphi) = ce^{\lambda\varphi}, \quad c, \lambda \in \mathbb{C}. \tag{1.29}$$

Substitution of (1.29) into (1.28) yields $\lambda = \pm\sqrt{-\lambda_2}$. Thus, for $\lambda_2 \neq 0$, the two-dimensional solution space of (1.28) is formed by all linear combinations of two obtained linearly independent solutions:

$$\Phi(\varphi) = c_1 e^{\sqrt{-\lambda_2}\varphi} + c_2 e^{-\sqrt{-\lambda_2}\varphi}, \quad c_1, c_2 \in \mathbb{C}.$$

In case of $\lambda_2 = 0$, the general solution of (1.28) is given by

$$\Phi(\varphi) = c_1 + c_2 x, \quad c_1, c_2 \in \mathbb{C}.$$

However, not all solutions of (1.28) are acceptable because of the requirement of 2π-periodicity. With the boundary conditions $\Phi(0) = \Phi(2\pi)$ and $\Phi'(0) = \Phi'(2\pi)$, non-trivial solutions exist only for $\lambda_2 = m^2, m \in \mathbb{N}_0$, and they are given by linear combinations of $e^{im\phi}$ and $e^{-im\phi}$.

In quantum mechanics it is customary to let $m \in \mathbb{Z}$ at this point and to express the solutions as linear combinations of

$$\Phi_m(\varphi) = e^{im\phi}, \quad m = \pm\sqrt{\lambda_2}.$$

m, which is referred as the *magnetic quantum number*, is subject to additional restrictions, as discussed below.

Equation (1.27) after the substitutions

$$x := \cos\vartheta, \quad y(x) := \Theta(\vartheta), \quad \lambda_2 = m^2$$

transforms into the general Legendre differential equation:

$$(x^2 - 1)y'' + 2xy' + \left(\frac{m^2}{1 - x^2} - \lambda_1\right)y = 0, \tag{1.30}$$

which can be solved using another handy substitution,

$$y(x) = (1 - x^2)^{|m|/2} z(x),$$

with a subsequent power series ansatz, as discussed in detail in [18]. One can show that well-behaved eigenfunctions exist only if

$$\lambda_1 = \ell(\ell + 1), \ \ell \in \mathbb{N}_0, \text{ and } |m| \leq \ell \tag{1.31}$$

(see, for example, [18] and references therein), whereas ℓ is referred as the *azimuthal*

TABLE 1.5: Some spherical harmonics (from [14], p. 302).

ℓ	m	$Y_{\ell,m}(\varphi,\vartheta)$
0	0	$\left(\dfrac{1}{4\pi}\right)^{1/2}$
1	0	$\left(\dfrac{3}{4\pi}\right)^{1/2}\cos\vartheta$
1	± 1	$\mp\left(\dfrac{3}{8\pi}\right)^{1/2}\sin\vartheta\, e^{\pm i\phi}$
2	0	$\left(\dfrac{5}{16\pi}\right)^{1/2}\left(3\cos^2\vartheta-1\right)$
2	± 1	$\mp\left(\dfrac{15}{8\pi}\right)^{1/2}\cos\vartheta\,\sin\vartheta\, e^{\pm i\phi}$
2	± 2	$\left(\dfrac{15}{32\pi}\right)^{1/2}\sin^2\vartheta\, e^{\pm 2i\phi}$

quantum number. The solutions of equation (1.30) under conditions given by (1.31) are *associated Legendre polynomials.* Therefore, the multiples of

$$\Theta_{\ell,m}(\vartheta) = \frac{(-1)^m}{2^\ell \ell!}\,(1-\cos^2\vartheta)^{m/2}\,\frac{d^{\ell+m}(\cos^2\vartheta-1)^\ell}{(d\cos\vartheta)^{\ell+m}} \tag{1.32}$$

are the only acceptable solutions of (1.27).

The normalized products of $\Phi_m(\varphi)$ and $\Theta_{\ell,m}(\vartheta)$ are called *spherical harmonics*:

$$Y_{\ell,m}(\varphi,\vartheta) := \sqrt{\frac{2\ell+1}{4\pi}\frac{(\ell-m)!}{(\ell+m)!}}\,\Phi_m(\varphi)\Theta_{\ell,m}.$$

Some important examples of spherical harmonics are listed in Table 1.5.

The radial equation (1.26) with a substitution

$$a := \frac{4\pi\epsilon_0\hbar^2}{mq_e^2},$$

after multiplication by $R(r)/r^2$ and replacement of λ_1 by $\ell(\ell+1)$ becomes:

$$R''(r) + \frac{2}{r}R'(r) + \left(\frac{2mE}{\hbar^2} + \frac{2Z}{ar} - \frac{\ell(\ell+1)}{r^2}\right)R(r) = 0. \tag{1.33}$$

The solutions of equation (1.33) are obtained by examination of their asymptotic behavior combined with a power series ansatz. A detailed discussion on this topic

TABLE 1.6: Some radial wavefunctions of a hydrogen-like atom (from [14], p. 324).

Orbital	n	l	$R_{n,l}(r)$
$1s$	1	0	$2\left(\dfrac{Z}{a}\right)^{3/2} e^{-Zr/na}$
$2s$	2	0	$\dfrac{1}{\sqrt{2}}\left(\dfrac{Z}{a}\right)^{3/2}\left(1-\dfrac{Zr}{na}\right) e^{-Zr/na}$
$2p$	2	1	$\dfrac{1}{\sqrt{6}}\left(\dfrac{Z}{a}\right)^{5/2}\dfrac{r}{n} e^{-Zr/na}$
$3s$	3	0	$\dfrac{1}{9\sqrt{3}}\left(\dfrac{Z}{a}\right)^{3/2}\left(6-\dfrac{6Zr}{na}+\left(\dfrac{Zr}{na}\right)^2\right) e^{-Zr/na}$
$3p$	3	1	$\dfrac{1}{9\sqrt{6}}\left(\dfrac{Z}{a}\right)^{5/2}\left(4-\dfrac{Zr}{na}\right)\dfrac{2r}{n} e^{-Zr/na}$
$3d$	3	2	$\dfrac{1}{9\sqrt{30}}\left(\dfrac{Z}{a}\right)^{7/2}\left(\dfrac{2r}{n}\right)^2 e^{-Zr/na}$

can be found, for example, in [18]. The normalized acceptable solutions, sometimes
called *radial wavefunctions*, are given by

$$R_{n,\ell}(r) = \sqrt{\left(\frac{2Z}{na}\right)^3 \frac{(n-\ell-1)!}{2n(n+\ell)!}} \left(\frac{2Zr}{na}\right)^\ell e^{-Zr/na} L_{n-\ell-1,2\ell+1}\left(\frac{2Zr}{na}\right), \quad (1.34)$$

where

$$n \in \mathbb{N}, \; n > \ell,$$

is the *principal quantum number*, and $L_{n,k}$ are *associated Laguerre polynomials*,
which can be defined, for example, as follows:

$$L_{n,k}(x) = \sum_{i=0}^{n} (-1)^i \frac{(n+k)!}{(n-i)!(k+i)!} \frac{x^i}{i!}. \quad (1.35)$$

Some important radial wavefunctions are listed in Table 1.6.

The only acceptable energy levels are:

$$E_n = -\frac{Z^2 m q_e^4}{32(\pi\epsilon_0 \hbar n)^2}.$$

Thus, they depend on n and not on the other quantum numbers. The ground state
corresponds to $n = 1$. As n increases, the separation of energy levels becomes
smaller, approaching continuous energy spectrum typical for unbound electrons.

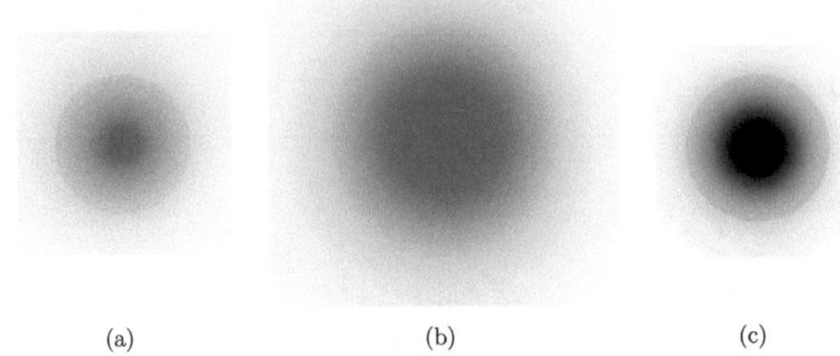

(a)   (b)   (c)

FIG. 1.8: $1s$-orbital (a, b) and the corresponding electron density (c) for the nuclear charge $1\,|e^-|$. The green shading of the 10%-opacity outlines the van der Waals sphere (see Subsection 1.5.4) of a hydrogen atom, centered in the nucleus. The opacity of the red and black shading in each depicted layer corresponds respectively to the value of the wavefunction and the electron density. (a) shows a slice through the atom center, (b) and (c) show seven layers with a 0.2-Å separation, such that the central layer passes through the nucleus.

One-electron wavefunctions that correspond to energy level E_n are linear combinations of

$$\Psi_{n,\ell,m} = R_{n,\ell}(r)Y_{\ell,m}(\varphi,\vartheta), \quad \ell = \overline{0, n-1}, \; m = \overline{-\ell, \ell} \tag{1.36}$$

for the given $n \in \mathbb{N}$. They are called *atomic orbitals*.

There are, in principle, infinitely many normalized linear combinations of the basis wavefunctions. However, it is customary to distinguish some standard atomic orbitals, a few of which are described below. Many other normalized linear combinations correspond to the same solutions rotated in space. Important exceptions are discussed later in this subsection.

To determine the state of an electron, one has to specify, along with the occupied orbital, the electron's *spin*, which is the intrinsic angular momentum of the electron. A spin of an electron is characterized by its *spin magnetic quantum number*, which can take only two values, $1/2$ and $-1/2$. The corresponding states are often referred as *up* (also ↑ or α) and *down* (also ↓ or β) spin respectively.

According to the *Pauli exclusion principle*, one orbital can be occupied by at maximum two electrons. Moreover, a double occupation of an orbital is only possible if the two electrons have *paired* (↑↓) spins. This statement, often encountered in chemical literature in either this or a similar form, in fact implies that the occupied orbitals must be orthogonal. Therefore, the number of orbitals that can be filled for each energy level is given by the number of basis functions spanning the solution space for this energy.

For the lowest energy level E_1, the solution space is one-dimensional, with a basis given by $\Psi_{1,0,0}$. All other normalized solutions give raise to the same electron

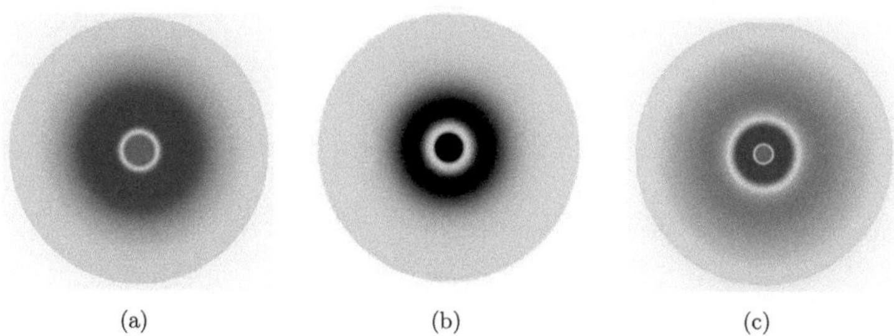

(a)   (b)   (c)

FIG. 1.9: Slices through the atom center, showing the $2s$-orbital (a), the corresponding electron density (b) for the nuclear charge $6\,|e^-|$, and the $3s$-orbital for the nuclear charge $11\,|e^-|$(c). Areas with the red and blue shading correspond to the regions where the wave function has positive and negative values respectively. The black shading outlines the electron density. The opacity of the shading corresponds to the absolute value of the visualized function in the depicted slice. For size comparison, a van der Waals sphere of a hydrogen atom, centered in the nucleus, is shaded by green of the 10%-opacity.

density. The wavefunction $\Psi_{1,0,0}$ is called the *1s-orbital* and also denoted Ψ_{1s} further in the text. It is shown together with the corresponding electron density $|\Psi_{1s}|^2$ in Figure 1.8. For a size reference, a van der Waals sphere of a hydrogen atom is shaded by light green.

Four standard orbitals that constitute a real orthonormal basis for the next energy level, E_2, are:

$$\Psi_{2s} := \Psi_{2,0,0} = R_{2,0}(r)\,\frac{1}{2\sqrt{\pi}} = \frac{1}{2\sqrt{2\pi}}\left(\frac{Z}{a}\right)^{3/2}\left(1 - \frac{Zr}{na}\right)e^{-Zr/na},$$

$$\Psi_{2p_x} := \frac{1}{\sqrt{2}}\left(\Psi_{2,1,-1} - \Psi_{2,1,1}\right) = R_{2,1}(r)\sqrt{\frac{3}{4\pi}}\frac{x}{r} = \frac{1}{2\sqrt{2\pi}}\left(\frac{Z}{a}\right)^{5/2}\frac{x}{n}e^{-Zr/na},$$

$$\Psi_{2p_y} := \frac{i}{\sqrt{2}}\left(\Psi_{2,1,-1} + \Psi_{2,1,1}\right) = R_{2,1}(r)\sqrt{\frac{3}{4\pi}}\frac{y}{r} = \frac{1}{2\sqrt{2\pi}}\left(\frac{Z}{a}\right)^{5/2}\frac{y}{n}e^{-Zr/na},$$

$$\Psi_{2p_z} := \Psi_{2,1,0} = R_{2,1}(r)\sqrt{\frac{3}{4\pi}}\frac{z}{r} = \frac{1}{2\sqrt{2\pi}}\left(\frac{Z}{a}\right)^{5/2}\frac{z}{n}e^{-Zr/na}.$$

Central slices of the $2s$-orbital and corresponding electron density for the nuclear charge $6\,|e^-|$ are shown in Figure 1.9, together with a van der Waals sphere of hydrogen, which is shaded by light green. Slices of the $2p$-orbitals for the nuclear charges $1\,|e^-|$, $6\,|e^-|$, and $16\,|e^-|$ are depicted in Figure 1.10 (a-c). Again, the light green shading outlines a van der Waals sphere of hydrogen, given for a size reference. One can see that for the hydrogenic nucleus charge the orbital is very diffuse and has a noticeable density extended over distances exceeding the van der Waals radius of hydrogen. This orbital is normally not occupied in a hydrogen atom. However, with increase of the nucleus charge, the orbital becomes more dense and compact.

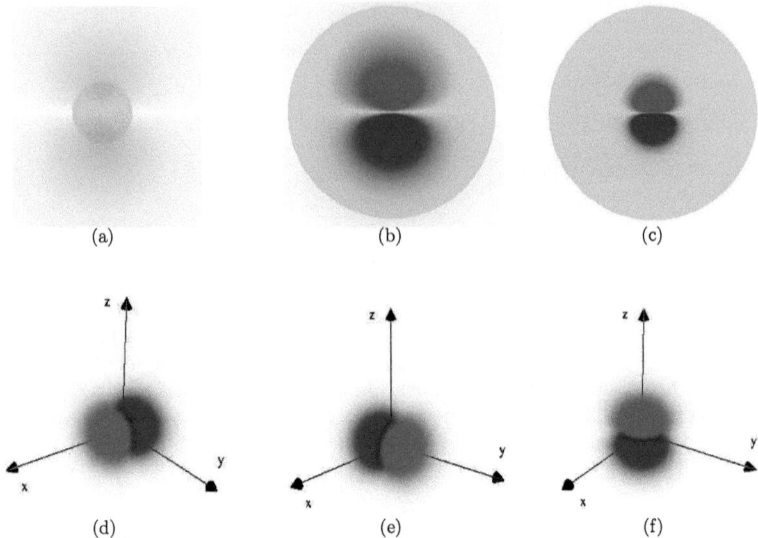

FIG. 1.10: Slices (a-c) through the atom center and 3D-reconstructions (d-f) by means of seven splices, showing real $2p$-orbitals. For comparison of $2p$-orbitals in cases of the nuclear charge $1\,|e^-|$(a), $6\,|e^-|$(b), and $16\,|e^-|$(c), a van der Waals sphere of a hydrogen atom, centered in the nucleus, is shaded by green of the 10%-opacity. As before, areas with the red and blue shading correspond to the regions where the wave function has positive and negative values respectively. (d) $2p_x$-orbital, (e) $2p_y$-orbital, (f) $2p_z$-orbital.

Figure 1.10 (d-f) shows the orientations of the standard $2p$-orbitals in the Cartesian coordinates. The real orbitals $2p_x$ and $2p_y$ are obtained by means of linear combinations of the complex-conjugated $\Psi_{2,1,-1}$ and $\Psi_{2,1,1}$. The same approach is used for construction of real bases for higher energy levels. Complex wavefunctions give raise to the real electron density, but they imply that the particle has a net momentum. For example, $\Psi_{2,1,-1}$ and $\Psi_{2,1,1}$ describe motions around the z axis (see, for example, [19] for details). Real orbitals correspond to standing waves with no net motion.

One distinguishes the following standard orbitals constituting a real orthonormal basis for E_3:

$$\Psi_{3s} := \Psi_{3,0,0} = \frac{1}{18\sqrt{3\pi}} \left(\frac{Z}{a}\right)^{3/2} \left(6 - \frac{6Zr}{na} + \left(\frac{Zr}{na}\right)^2\right) e^{-Zr/na},$$

$$\Psi_{3p_x} := \frac{1}{\sqrt{2}} \left(\Psi_{3,1,-1} - \Psi_{3,1,1}\right) = \frac{1}{9\sqrt{2\pi}} \left(\frac{Z}{a}\right)^{5/2} \left(4 - \frac{Zr}{na}\right) \frac{x}{n} e^{-Zr/na},$$

$$\Psi_{3p_y} := \frac{i}{\sqrt{2}} \left(\Psi_{3,1,-1} + \Psi_{3,1,1}\right) = \frac{1}{9\sqrt{2\pi}} \left(\frac{Z}{a}\right)^{5/2} \left(4 - \frac{Zr}{na}\right) \frac{y}{n} e^{-Zr/na},$$

$$\Psi_{3p_z} := \Psi_{3,1,0} = \frac{1}{9\sqrt{2\pi}} \left(\frac{Z}{a}\right)^{5/2} \left(4 - \frac{Zr}{na}\right) \frac{z}{n} e^{-Zr/na},$$

$$\Psi_{3d_{z^2}} := \Psi_{3,2,0} = \frac{1}{9\sqrt{6\pi}} \left(\frac{Z}{a}\right)^{7/2} \frac{2z^2 - x^2 - y^2}{n^2} e^{-Zr/na},$$

$$\Psi_{3d_{xy}} := \frac{i}{\sqrt{2}} \left(\Psi_{3,2,-2} - \Psi_{3,2,2}\right) = \frac{1}{9}\sqrt{\frac{2}{\pi}} \left(\frac{Z}{a}\right)^{7/2} \frac{xy}{n^2} e^{-Zr/na},$$

$$\Psi_{3d_{xz}} := \frac{1}{\sqrt{2}} \left(\Psi_{3,2,-1} - \Psi_{3,2,1}\right) = \frac{1}{9}\sqrt{\frac{2}{\pi}} \left(\frac{Z}{a}\right)^{7/2} \frac{xz}{n^2} e^{-Zr/na},$$

$$\Psi_{3d_{yz}} := \frac{i}{\sqrt{2}} \left(\Psi_{3,2,-1} + \Psi_{3,2,1}\right) = \frac{1}{9}\sqrt{\frac{2}{\pi}} \left(\frac{Z}{a}\right)^{7/2} \frac{yz}{n^2} e^{-Zr/na},$$

$$\Psi_{3d_{x^2-y^2}} := \frac{1}{\sqrt{2}} \left(\Psi_{3,2,-2} + \Psi_{3,2,2}\right) = \frac{1}{9\sqrt{2\pi}} \left(\frac{Z}{a}\right)^{7/2} \frac{x^2 - y^2}{n^2} e^{-Zr/na}.$$

Central slices of the $3s$-, $3p$-, and $3d$-orbitals are depicted in Figures 1.9 (c), 1.11 (a), and 1.11 (b,c) respectively. Figures 1.11 (d-i) show the distribution and orientation of the $3p_z$- and described standard $3d$-orbitals in space.

The orbitals related to the same principal quantum number n are said to constitute a single *shell* of the atom, while the orbitals associated with different azimuthal quantum numbers ℓ belong to different *subshells*, which are referred by letters s, p, d, and f for ℓ equal to 0, 1, 2, and 3 respectively.

Easy to see that the described standard np-orbitals for a given n have the same shape, but different symmetry axes, which coincide with the x, y, and z axis of the Cartesian coordinates respectively. Any normalized wavefunction obtained by a real linear combination of those p-orbitals has again the same shape rotated in space. For example, let

$$\Psi_{np} := N(a\Psi_{np_x} + b\Psi_{np_y} + c\Psi_{np_z}), \tag{1.37}$$

where a, b, $c \in \mathbb{R}$, and N is the normalization constant. Since Ψ_{np_x}, Ψ_{np_y}, and Ψ_{np_z} are normalized and orthogonal, we obtain:

$$N = \frac{1}{\sqrt{a^2 + b^2 + c^2}}.$$

Equation (1.37) after substitution of the appropriate spherical harmonics can be rearranged as:

$$\Psi_{np} = R_{n,1}\sqrt{\frac{3}{4\pi}} \frac{N(ax + by + cz)}{r}.$$

Thus, the new orbital has the same structure with the symmetry axis given by the unit vector $N(a, b, c)^{\mathrm{T}}$.

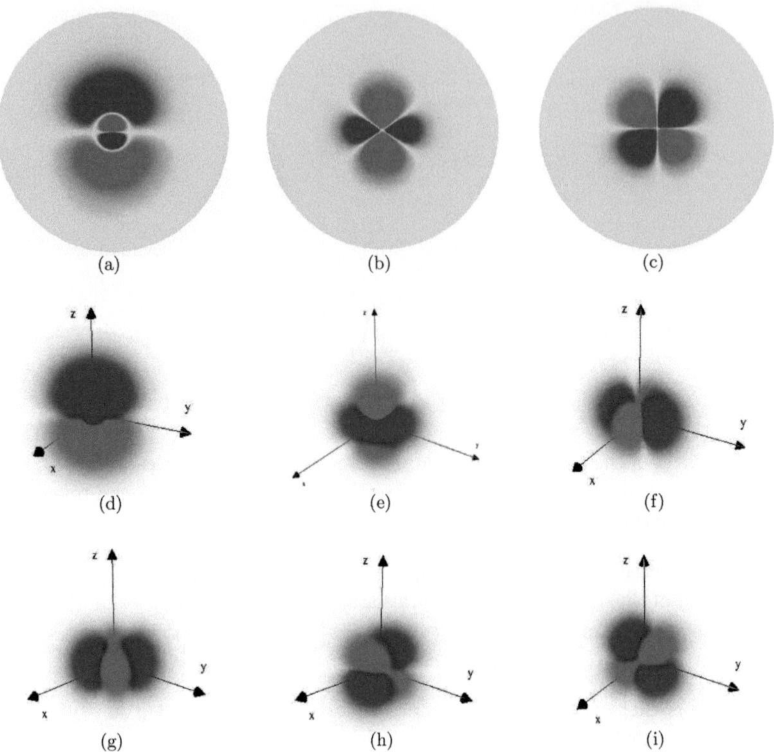

FIG. 1.11: Slices through the atom center (a-c) and 3D-reconstructions by means of seven splices (d-i), showing real $3p$- (a,d) and $3d$-orbitals. The green shading of the 10%-opacity in (a-c) outlines the van der Waals sphere of a hydrogen atom, centered in the nucleus. (a,d) the $3p_z$-orbital for the nuclear charge $16\,|e^-|$; (b,e) the $3d_{z^2}$-orbital for the nuclear charge $21\,|e^-|$; (c) the other $3d$-orbitals for the nuclear charge $21\,|e^-|$; (f) the $3d_{x^2-y^2}$-orbital; (g) the $3d_{xy}$-orbital; (h) the $3d_{xz}$-orbital; (i) the $3d_{yz}$-orbital.

If we combine the s- and p-orbitals of the same shell, we obtain hybrid orbitals (see Figure 1.12). In these cases the interplay of the constructive and destructive interference of waves can give raise to new shapes (see also Figure 1.13), depending on the proportion of the s- and p-orbitals. That is, various combinations with the same proportion of the s-orbital give an identical shape with different orientations in space.

Eligible linear combinations may also include d-orbitals, but by reasons explained later they are of no direct relevance for protein modeling, therefore we shall omit them in our discussion.

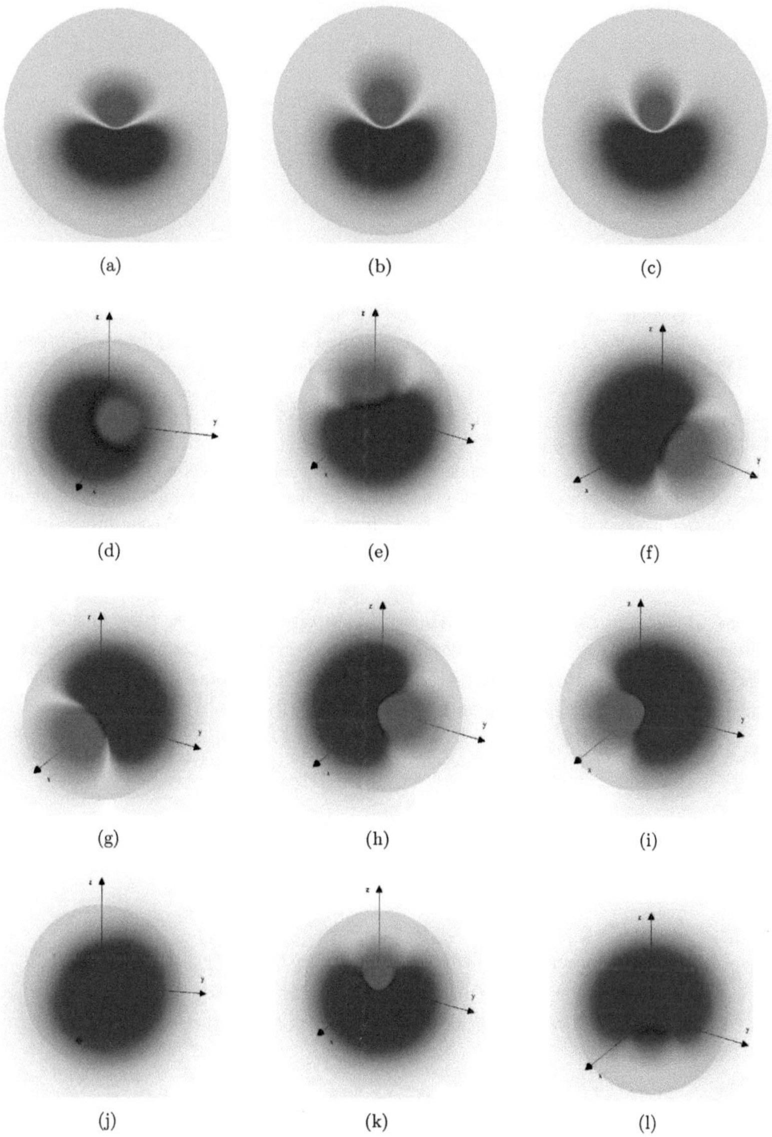

FIG. 1.12: Slices through the atom center (a-c) and 3D-reconstructions by means of seven splices (d-l), showing hybridized $2sp^3$- (a,d-g), $2sp^2$- (b,h-j) and $2sp$-orbitals (c,k,l) for the nuclear charge $6\,|e^-|$. The green shading of the 10%-opacity as before outlines the van der Waals sphere of a hydrogen atom, centered in the nucleus.

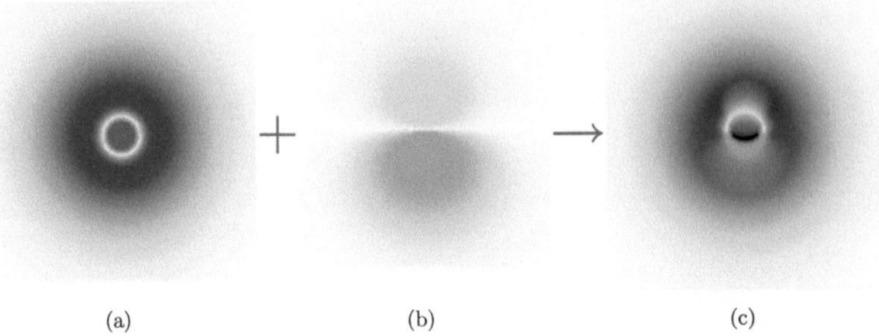

(a)   (b)   (c)

FIG. 1.13: Formation of a hybridized $2sp$-orbital by interference of a $2s$- and $2p$-orbital. (a) a central slice of a $2s$-orbital scaled by $1/\sqrt{2}$. Red and blue shading denote the regions where the wavefunction has a positive or a negative value respectively. (b) a central slice of a $2p$-orbital scaled by $1/\sqrt{2}$. The regions of positive or negative values of the wavefunction are shaded by yellow and cyan respectively. (c) the interference of the orbitals depicted in (a) and (b). The scaling factor $1/\sqrt{2}$ is the normalization constant for an sp-hybrid. Red and yellow areas correspond to the regions where the resulting wavefunction has a positive value, while blue and cyan colors relate to the regions of a negative value. Neutral gray and black colors correspond to the regions where the wavefunctions cancel each other through the destructive interference (see also Figure 1.12 (c)).

In general, we are interested in orthonormal orbital sets, in a view of the Pauli exclusion principle. One set, typically discussed in the context of valence bond theory, in an explanation of the tetrahedral shape of methane, consists of four sp^3 hybrid orbitals, to which the s- and p-orbitals contribute in a ratio 1:3:

$$\Psi^{[1]}_{n\,sp^3} := \frac{1}{2}\left(\Psi_{ns} + \Psi_{np_x} + \Psi_{np_y} + \Psi_{np_z}\right),$$

$$\Psi^{[2]}_{n\,sp^3} := \frac{1}{2}\left(\Psi_{ns} - \Psi_{np_x} - \Psi_{np_y} + \Psi_{np_z}\right),$$

$$\Psi^{[3]}_{n\,sp^3} := \frac{1}{2}\left(\Psi_{ns} - \Psi_{np_x} + \Psi_{np_y} - \Psi_{np_z}\right),$$

$$\Psi^{[4]}_{n\,sp^3} := \frac{1}{2}\left(\Psi_{ns} + \Psi_{np_x} - \Psi_{np_y} - \Psi_{np_z}\right).$$

Their major lobes point to the directions of corners of a regular tetrahedron (see Figure 1.12 (d-g)). Two other sets that we shall note are three sp^2-hybrids,

$$\Psi^{[1]}_{n\,sp^2} := \frac{1}{\sqrt{3}}\left(\Psi_{ns} + \sqrt{2}\Psi_{np_y}\right),$$

$$\Psi^{[2]}_{n\,sp^2} := \frac{1}{\sqrt{3}}\left(\Psi_{ns} + \sqrt{\frac{3}{2}}\Psi_{np_x} - \frac{1}{\sqrt{2}}\Psi_{np_y}\right),$$

$$\Psi^{[3]}_{n\,sp^2} := \frac{1}{\sqrt{3}}\left(\Psi_{ns} - \sqrt{\frac{3}{2}}\Psi_{np_x} - \frac{1}{\sqrt{2}}\Psi_{np_y}\right),$$

combined with the standard np_z-orbital, and two sp-hybrids,

$$\Psi^{[1]}_{n\,sp} = \frac{1}{\sqrt{2}}\left(\Psi_{n\,s} + \Psi_{n\,p_z}\right),$$

$$\Psi^{[2]}_{n\,sp} = \frac{1}{\sqrt{2}}\left(\Psi_{n\,s} - \Psi_{n\,p_z}\right),$$

combined with the remaining standard np-orbitals, $\Psi_{n\,p_x}$ and $\Psi_{n\,p_y}$. The major lobes of the described sp^2-hybrids point to the directions at $120°$ to each other (see Figure 1.12 (h-j)), in the plane orthogonal to the symmetry axis of the remaining p-orbital. The major lobes of the two sp-hybrids point to the opposite directions (see Figure 1.12 (k,l)), which are orthogonal to the symmetry axes of the remaining p-orbitals in the set. Unlimited number of equivalent sets can be obtained from other linear combinations, yielding the same structures rotated in space. We shall return to them in discussion of covalent bonding.

In principle, one can obtain other hybrids with different percentage of s-character, which would be also eligible orbitals for a hydrogen-like atom. However, in systems with many electrons (and possibly in presence of other nuclei), the interactions of the latter come into play, which perturb the described solutions, giving a favor to one or another arrangement.

The probability to find an electron occupying a certain orbital $\Psi_{n,\ell,m}$ in a layer of thickness Δr, separated by a distance r from the nucleus, is given by

$$\int\limits_{r}^{r+\Delta r}\int\limits_{0}^{\pi}\int\limits_{0}^{2\pi} R^2_{n,\ell}(\tilde{r})|Y_{\ell,m}(\varphi,\vartheta)|^2\tilde{r}^2\sin\vartheta\,d\varphi\,d\vartheta\,d\tilde{r} = \int\limits_{r}^{r+\Delta r} R^2_{n,\ell}(\tilde{r})\tilde{r}^2\,d\tilde{r}.$$

Thus,

$$P_{n,\ell}(r) := r^2 R^2_{n,\ell}(r)$$

is the probability density to observe an electron at a certain distance from the nucleus.

If we compare radial distribution functions of s and p orbitals, we see that an electron occupying an s-orbital is likely to be found closer to the nucleus than a p-electron in the same shell, while a d-electron is less tightly bound to the nucleus. Therefore, in many-electron atoms s-electrons experience less shielding by the electrons of the inner shells, and p-electrons, in turn, are less shielded from the nucleus then d-electrons. Hence, the energies of the subshells in a single shell are, in fact, not equal in atoms with many electrons. This can be observed in atomic spectra, which also reflect energy differences arising from spin correlation effects and spin-orbit coupling, not discussed here.

Existing analytical approximations for orbitals in many-electron atoms give orbital shapes similar to those of the described standard orbitals in hydrogen-like atoms.

TABLE 1.7: Ground-state configurations of atoms encountered in proteins.

Atom type	Nuclear charge	Ground-state configuration
H	1	$1s$
C	6	$1s^2 2s^2 2p^2$
N	7	$1s^2 2s^2 2p^3$
O	8	$1s^2 2s^2 2p^4$
S	16	$1s^2 2s^2 2p^6 3s^2 3p^4$

Only the hydrogenic orbital exponents have to be adjusted to account for the shielding effect [20]. Moreover, images obtained by high-resolution scanning tunneling microscopy resemble atomic orbitals derived analytically for hydrogen-like atoms [21].

In accordance with the *building-up principle* (often referred as the *Aufbau principle*), the ground state configuration of an atom with the nuclear charge Z is predicted as follows: Z electrons are placed subsequently into the orbitals $1s$, $2s$, $2p$, $3s$, $3p$, $4s$, $3d$, and so on, up to two electrons with paired spins per an orbital, and taking into account the degeneracy of each subshell. One shall note that the $3d$-subshell is usually filled after the $4s$-subshell. This can be explained by a stronger repulsion between electrons in a d-subshell [14].

The ground states of atoms that are encountered in proteins are given in Table 1.7.

1.5.2 COVALENT BONDING

The ability of an atom to form chemical bonds depends on the number of its *valence electrons*, i.e. the electrons in the outermost shell of the atom in its ground state. When the outer shell is filled by the maximal number of electrons, atoms tend to be chemically inert. By contrast, atoms with electron vacancies in the outer shell can initiate spin coupling with electrons of other atoms and thereby form joint orbitals with shared electron pairs. This results in *covalent bonding* of the participating atoms.

According to valence-bond theory, a σ-*bond* originates from an overlap of orbitals, the symmetry axes of which coincide with the axis of the bond. It is implied that s-orbitals among others are capable of forming σ-bonds, since for s-orbitals any axis through the nucleus center is a symmetry axis.

π-*bonds*, which are weaker than σ-bonds, may be formed additionally from approaching p-orbitals with the symmetry axes orthogonal to the line connecting atom centers. Thus, π-bonding manifests itself in double and triple bonds between atoms.

The only orbital occupied in a ground state of a hydrogen atom is the $1s$-orbital, therefore hydrogen is only capable of σ-bonding. The ground-state configuration of a carbon atom assumes that electrons on its $2s$-orbital have paired spins, there-

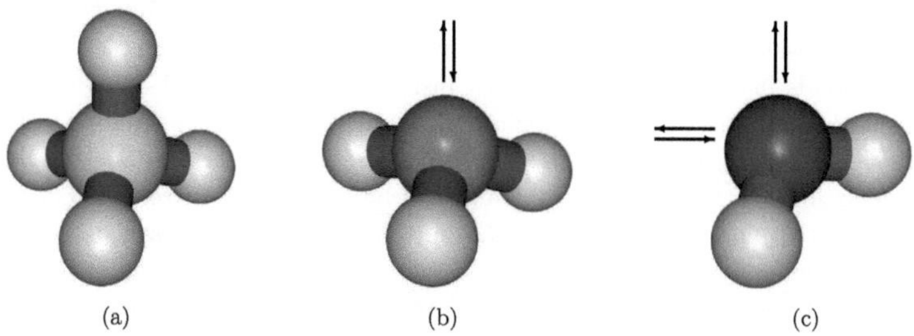

(a)   (b)   (c)

Fig. 1.14: Approximate structure of methane, ammonia and water molecules. Arrows designate lone pairs.

fore carbon should be capable of forming only two bonds, via spin pairing of the remaining two valence electrons occupying different $2p$-orbitals (see Table 1.7). To explain the tetravalence of carbon, valence bond theory suggests that one of the electrons from the $2s$-subshell is excited to the vacant $2p$-orbital, with a subsequent formation of four sp^3-orbitals, each occupied by a valent electron capable of spin pairing.

Thus, according to this interpretation, four σ-bonds, pointing to the vertices of a regular tetrahedron, are formed by an sp^3-hybridized carbon (see Figure 1.14 (a)). If, however, the number of reactive atoms is not sufficient for forming four σ-bonds, an sp^2 configuration of the valence shell may be preferable. In this case, the p-orbital that does not participate in hybridization and formation of σ-bonds can be involved in π-bonding. As a result, the axes of the σ-bonds in an sp^2-hybridized carbon point to the vertices of a regular triangle centered at the nucleus, while the orbital of the π-bond is located at the two sides of the triangle plane.

The ground-state configuration of a nitrogen atom (see Table 1.7) suggests that its two valence electrons in the $2s$-subshell have paired spins, while three electrons in the $2p$-subshell occupy three different p-orbitals. Therefore, nitrogen in its ground state is capable of forming three covalent bonds. Nevertheless, its valence shell also adopts nearly an sp^3 configuration when only single bonds are formed (see Figure 1.14 (b)) and an sp^2-like arrangement, if the atom is involved in π-bonding. A pair of valence electrons that does not participate in covalent bonding is called a *lone pair*.

Likewise, oxygen and sulfur atoms have only two valence electrons with unpaired spins, accommodated in the $2p$- and $3p$- subshell respectively, while the other four valence electrons share in couples two other orbitals of the outer shell. Thus, oxygen and sulfur form only two covalent bonds. Again, in a case of only single bonds all orbitals of their valence shells hybridize (see Figure 1.14 (c)), while if a double bond is formed, one of the p-orbitals is involved in π-bonding, and the remaining ones hybridize.

The presence of a lone pair in an atom may decrease angles between bonds, owing to stronger repulsion of the shared electrons by the lone pair. For example, the angle between bonds in water is about $104.5°$, instead of the predicted angle of $109.5°$ between sp^3-hybridized orbitals. Some authors relate it to an unequal proportion of the s-character in the formed hybrid orbitals. Formally, non-standard hybrids are also eligible hydrogenic orbitals, which can be preferred if there is no complete symmetry to rationalize an equivalent proportion. However, the formed bonds are not simple overlaps of standard atomic orbitals with spin coupling, as was initially stated by valence-bond theory. According to the more modern molecular orbital theory, each bond is a new orbital equally occupied by two electrons with paired spins. Therefore, exact quantification of the s- and p-proportion in a configuration that gave raise to the new orbitals is of no direct importance, and we shall adopt a simple terminology, distinguishing only between the discrete hybridization states described above.

Since the symmetry axes of p-orbitals involved in π-bonding must lay in one plane, twisting about such bonds is prohibited. Another important issue is that π-bonds tend to give raise to *resonance effects*, characterized by delocalization of electrons. This has an impact on bond lengths and may result in hindrance of twisting about some formally single bonds. For instance, the nitrogen in amide groups adopts an sp^2-like arrangement, because its lone pair is attracted by the carbon to an orbital resembling a π-bond, while the actual π-bond between the carbon and the oxygen is, in turn, shifted to the latter more electronegative* element. This has a direct consequence for a protein structure, see Subsection 1.6.1. Similar electron delocalization happens in carboxyl ions and in the side chain of arginine. Aromatic rings, present in amino acid side chains, are well-known examples of resonance structures. Besides, a lone pair of oxygen in the side chain of a tyrosine molecule is partially involved into the π-system of the aromatic ring. Therefore, the hydroxyl group of the tyrosine side chain is likely to be found in the same plane with the ring atoms.

1.5.3   ELECTROSTATIC INTERACTIONS

When a covalent bond is formed by two atoms of different types, the electron density related to shared electrons is not distributed evenly between the two nuclei. Depending on the properties of the atomic electron shells and on the nuclear charges, it is energetically favorable for the shared electron pair to spend more time near one or another nucleus. The neighborhood or, in particular, bonds with other atoms can result in a shift of the electron density even in covalent bonds between nuclei of the same type.

Redistribution of the electron density in a molecule results in accumulation of negative *partial charges* at certain atoms, while the other atoms obtain corresponding positive partial charges (see Figure 1.15). All partial charges add up to zero in a

---

*For a discussion of the electronegativity see the next section.

FIG. 1.15: Putative* distribution of partial charges in a molecule of arginine. Red and blue tincture denote respectively positive and negative charge. Full intensity of red or blue would imply a charge with an absolute value of $1\,|e^-|$.

neutral molecule. However, an uneven spatial distribution of charges has an effect that many molecules may posses an *electric dipole moment*, which can be detected experimentally.

If the charge is distributed in the volume $\Omega \subset \mathbb{R}^3$, the electric dipole moment is defined as

$$\vec{\mu}(\vec{r}_0) := \int_{\Omega} q(\vec{r})(\vec{r} - \vec{r}_0)d\vec{r} \tag{1.38}$$

where $q(\vec{r})$ is the charge density in $\vec{r} \in \Omega$, and $\vec{r}_0$ is the observation point. For a set of N point charges, their density can be expressed using Dirac delta functions, and equation (1.38) transforms into

$$\vec{\mu}(\vec{r}_0) = \sum_{i=1}^{N} q_i(\vec{r}_i - \vec{r}_0), \tag{1.39}$$

where q_i is the charge located in the point $\vec{r}_i$, $i = \overline{1, N}$.

The energy $G_{ij}^{[e]}$ of the interaction between two point charges q_i and q_j ($i \neq j$) is described by Coulomb's law with a correction for the dielectric screening:

$$G_{ij}^{[e]} = \frac{q_i q_j}{4\pi\epsilon_0\epsilon_r r}, \tag{1.40}$$

where ϵ_r is the relative permittivity of the medium and r is the distance between charges. The vacuum permittivity ϵ_0 was already employed in Subsection 1.5.1. Strictly speaking, $G_{ij}^{[e]}$ is a free energy[†] [1], since it involves entropy changes in the medium and depends on temperature.

The relative permittivity ϵ_r accounts for polarization of the medium. Being equal to 1 in vacuum, it raises to values between 2 and 4 in a protein core, and achieves 80 in water [1]. Presence of small ions abundant in cytosol further increases the screening effect.

---

*Charges are computed by SiViProf using the procedure described in Subsection 2.2.3.
[†]Unless the interaction occurs in vacuum.

Equation (1.40) implies a homogeneous medium. However, charged atoms are usually located at the protein surface*, apart from those in peptide groups, which are also inevitably present in the protein core. The effective ϵ_r for the interaction of protein charges is variable and depends not only on their immediate surrounding. In the most typical cases, an appropriate ϵ_r for the evaluation of the interaction between two surface charges is probably 40, although it can achieve about 200, if the charges are located at the opposite sites of a protein globule [1]. The screening effect is substantial even at distances when no water molecules can be placed between interacting charges. However, at such distances quantum mechanical effects dominate.

The ability of an atom to attract an electron pair in a covalent bond is often evaluated in terms of *electronegativity*. This concept was first introduced by L. Pauling, who proposed an electronegativity scale based on bond dissociation energies (see, for example, [14], p. 379). An alternative definition of electronegativity, which became widely known, was suggested by R. Mulliken [22]. It is expressed through the ionization energy I and the electron affinity E of an atom:

$$\chi := \frac{1}{2}(I + E).$$

This definition was further elaborated by J. Hinze [23–25], who suggested that electronegativity should be a characteristic of a specific orbital in a certain valence state. According to this definition, electronegativity $\chi^{[n]}$ of the n-th orbital is given by

$$\chi^{[n]} := \frac{1}{2}(I^{[n]} + E^{[n]}), \tag{1.41}$$

where $I^{[n]}$ is the corresponding ionization potential and $E^{[n]}$ is the electron affinity of this orbital.

Based on the latter definition, J. Gasteiger and M. Marsili [26, 27] developed a procedure for rapid computation of partial charges, which is utilized in this work and discussed in details in Subsection 2.2.3 of the next chapter.

1.5.4 VAN DER WAALS FORCES

When two non-bonded atoms come too close, electron shells start to repel each other in accordance to the Pauli exclusion principle. However, at larger distances even neutral atoms exert mutual attraction. Since the electrons in an atom constantly move, they produce an instantaneous dipole moment, which induces an antiparallel dipole moment in another atom. The correlated change of dipole moments results in attractive force. The energy of this interaction decays proportionally to the sixth degree of the separation between atom centers. Both repulsive and attractive interactions of this kind are termed collectively as *van der Waals interactions*.

---

*Strictly speaking, all atoms in the protein have certain partial charges. However, most of them are insignificant compared to those in polar or ionized amino acid side chains.

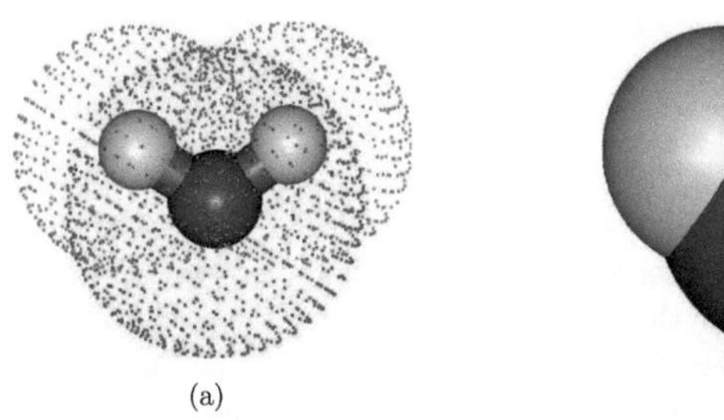

(a)   (b)

FIG. 1.16: Models of a water molecule: (a) ball-and-stick model, where van der Waals spheres of atoms are outlined by dots, (b) a space-filling model, where atoms are shown as van der Waals spheres.

Although the electron clouds do not have to be spherical, it is convenient to model atoms as hard spheres, assuming that potential energy rises abruptly when the separation between sphere centers becomes smaller than the sum of their radii, often referred as *van der Waals radii*. These radii are conventionally assumed to be a characteristic related to the type of the atom, but not to the atomic state. Spheres with van der Waals radii are sometimes briefly called the *van der Waals spheres* of atoms (see Figure 1.16). The filled union of the van der Waals spheres that are positioned in the atom centers constitutes the *molecular volume*. The surface of this union is called the *van der Waals envelope*, or *surface*, of the molecule.

The energy of the van der Waals interaction, which depends on the separation r between atom centers, is approximately described by the Lennard-Jones potential (Fig. 1.17):

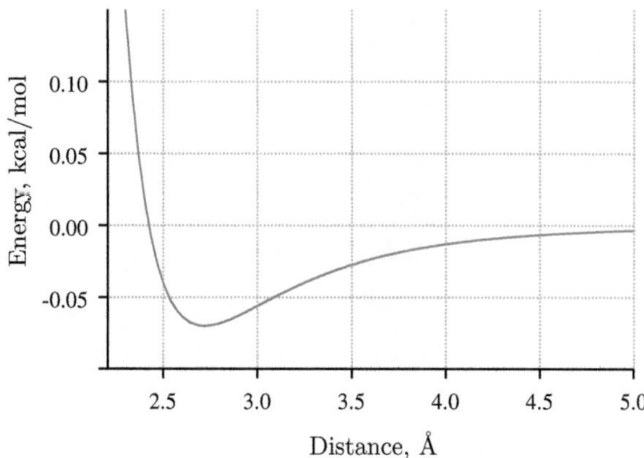

FIG. 1.17: Lennard-Jones potential of van der Waals interaction between an oxygen and a hydrogen atom.

$$U(r) = E_0 \left(\left(\frac{r_0}{r} \right)^{12} - 2 \left(\frac{r_0}{r} \right)^6 \right). \tag{1.42}$$

Here E_0 is the absolute value of the minimal energy and r_0 is the distance at which the minimum is achieved.

The first term in the parenthesis is responsible for a drastic increase in energy when the separation between atom centers becomes smaller then r_0. The second term dominates at distances larger than $r_0/\sqrt[6]{2}$. It stands for the attractive part of the interaction, and its r^{-6}-order decay can be derived from quantum mechanics [5]. The form of the repulsive term is chosen mainly for convenience of computation. It describes the behavior only qualitatively by preventing excessive convergence of atoms.

Since the absolute value of the interaction energy decays very fast with increasing distance, it is usual to use a certain cutoff for evaluation of the van der Waals energy in computations involving a large number of atoms. That is, if the distance between two atoms exceeds the cutoff value, their interaction energy is not computed.

1.5.5   HYDROGEN BONDING

Along with disulfide bridges and van der Waals interactions, which were discussed in previous sections, an important factor stabilizing protein structure is *hydrogen bonding* (see Figure 1.18). Unlike disulfide bridges, which mostly occur in secretory proteins, hydrogen bonds are numerous in almost any native protein.

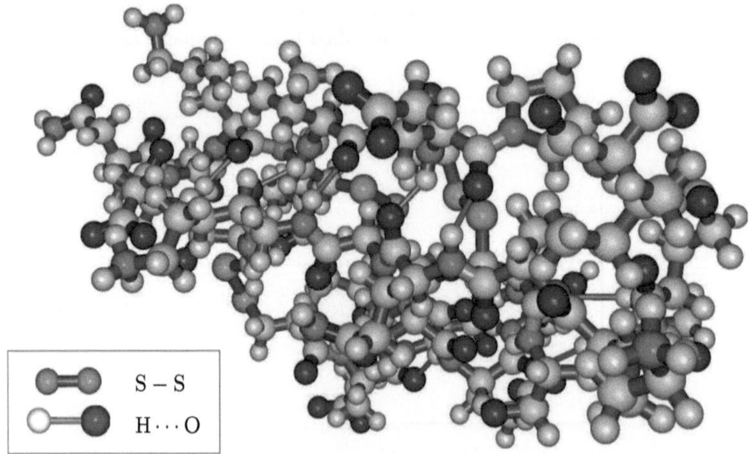

FIG. 1.18: The solution structure of the scorpion peptide P01, deduced by nuclear magnetic resonance. Atomic coordinates are obtained from RCSB Protein Data Bank, record 1ACW by E. Blanc *et al.* [28]. Disulfide (S – S) and hydrogen (i.e. H · · · O) bonds, holding chain fragments together, are reconstructed by SIVIPROF.

Having the electrostatic nature, hydrogen bonding is significantly weaker then covalent, but stronger then an attractive van der Waals interaction. Water is liquid due to polarity of its molecules and the resulting hydrogen bonding, which determine some peculiar effects, discussed further in Subsections 1.7.1-1.7.2. Hydrogen sulfide (H_2S) has a similar composition, only the oxygen atom is replaced by a heavier sulfur atom. Nevertheless, H_2S is gaseous, because sulfur is less electronegative and less capable of forming hydrogen bonds. Typical energy of a hydrogen bond in water can be estimated to about 5 kcal [1].

The electron shell of a hydrogen atom consists of only one electron. When a covalent bond with a strongly electronegative atom, such as oxygen or nitrogen, is formed, the electron density is shifted to that atom, leaving the hydrogen nucleus largely uncovered. Such a pair is a potential *donor of a hydrogen bond.* If another electronegative atom with a lone or π-bonding pair of electrons, being a potential *acceptor of the hydrogen bond,* approaches from the other side, it interacts with the hydrogen nucleus and pushes the electron density from the hydrogen even further away. As a result, the acceptor of the hydrogen bond can come as close to the electronegative atom of the donor group, as if there were no hydrogen between them (see Figure 1.19 (a)).

Hydrogen bonds are rather sensitive to the orientation of the covalent bond from the donor group. The latter has to be directed at the acceptor, and deviations usually do not exceed twenty to thirty degrees [1]. A donor group can donate only one hydrogen bond, while the maximal number of accepted hydrogen bonds is determined by the number of the unshared or π-bonding electron pairs in the acceptor. In most favorable configurations, the direction of the covalent bond in the donor group coincides with the supposed axis either of a lone pair, or of the π-bond, if the acceptor has no non-bonding electron pairs [29]. Thus, in a water molecule, the sp^3-hybridized oxygen has two covalent bonds with hydrogens, which are capable of donating altogether two hydrogen bonds. Besides, it has two unshared

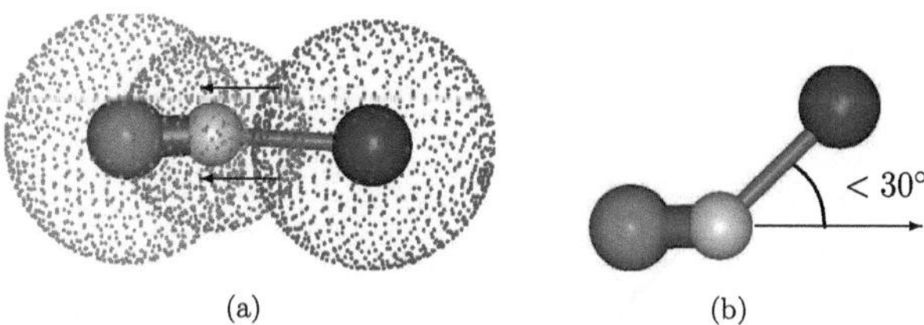

FIG. 1.19: Geometric features of hydrogen bonds. Donor group is on the left-hand side, acceptor – on the right. (a) Electron density is shifted from the hydrogen atom to the electronegative atom of the donor group, therefore the electronegative acceptor can approach the hydrogen nucleus. (b) Hydrogen bonds are sensitive to the direction of the covalent bond from the donor group.

electron pairs and therefore is able to accept two other hydrogen bonds at the angle about 109° to each other.

1.6 THREE-DIMENSIONAL STRUCTURE OF PROTEINS

Bond lengths and angles between any two bonds of the same atom, termed *bond angles*, are determined by properties of electron shells and almost do not change during the process of protein folding. In fact, bonds vibrate with very high frequency [14], but appear rigid at the folding time scale. Quantum mechanical computations show that bonds vibrating with the lowest possible energy are characterized by the largest probability to be found with zero deviation from their equilibrium length [14]. At the same time, bond vibrations can not be excited to higher levels at physiological conditions. Bond angles also oscillate and, compared to bond lengths, appear less rigid, since their vibrational frequency can change at normal temperature [1]. Nevertheless, analysis of native protein structures indicates that bond angles make only minor contributions to the protein flexibility [1]. Standard deviations less then 0.2 Å for bond lengths and approximately 2° for bond angles were reported for biological macromolecular structures (see the review [5] and references therein).

Thus, typical bond lengths appear fixed roughly between 1 and 2 Å, while bond angles usually remain about either 109° or 120°, depending on the atom hybridization state. However, nascent protein chain remains flexible, because twisting about single bonds with displacement of atom groups relative to each other is allowed. Neglecting the fluctuations in bond angles and lengths, one can prescribe the polypeptide conformation by giving the degree of twisting around each bond connecting non-hydrogen atoms. For this purpose dihedral, or torsion, angles are used.

1.6.1 DIHEDRAL ANGLES

Let A_i, A_j, A_k, A_l be the centers of four consequently connected atoms, and p_{ijk}, p_{jkl} be the planes through the points A_i, A_j, A_k and A_j, A_k, A_l respectively.

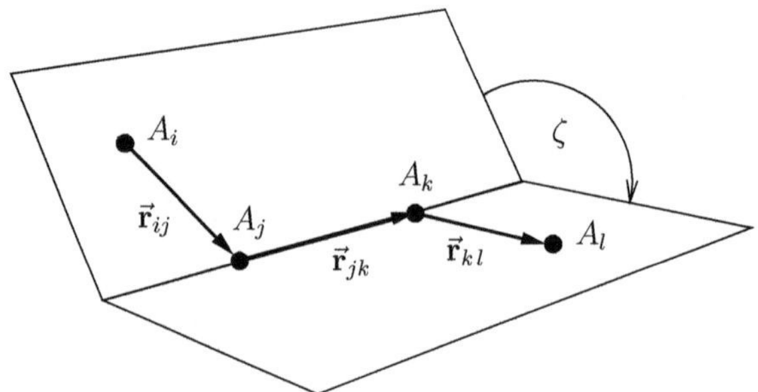

FIG. 1.20: Dihedral angle $\zeta = \angle(A_i - A_j - A_k - A_l)$.

In application to molecules, the *dihedral angle* $\angle(A_i - A_j - A_k - A_l) \in (-\pi, \pi]$ is the angle between the planes p_{ijk} and p_{jkl} that is measured as shown in Figure 1.20. Let $\vec{r}_{ij}$, $\vec{r}_{jk}$, $\vec{r}_{kl}$ be vectors pointing from A_i to A_j, from A_j to A_k, and from A_k to A_l respectively, such that $\vec{r}_{ij} \not\parallel \vec{r}_{jk}$ and $\vec{r}_{jk} \not\parallel \vec{r}_{kl}$. Let $\vec{n}_{ijk}$, $\vec{n}_{jkl}$ be normals of planes p_{ijk} and p_{jkl}, directed in a way that the vectors $\vec{r}_{ij}$, $\vec{r}_{jk}$, $\vec{n}_{ijk}$ and $\vec{r}_{jk}$, $\vec{r}_{kl}$, $\vec{n}_{jkl}$ build right-handed triples. Then the dihedral angle $\angle(A_i - A_j - A_k - A_l)$ can be computed as an angle between the normals $\vec{n}_{ijk}$ and $\vec{n}_{jkl}$ by means of equations* (1.44)-(1.46):

$$\vec{n}_{ijk} = \vec{r}_{ij} \times \vec{r}_{jk}, \tag{1.44}$$

$$\vec{n}_{jkl} = \vec{r}_{jk} \times \vec{r}_{kl}, \tag{1.45}$$

$$\angle(A_i - A_j - A_k - A_l) = \text{sign}(\vec{n}_{jkl} \cdot \vec{r}_{ij}) \arccos \frac{\vec{n}_{ijk} \cdot \vec{n}_{jkl}}{\|\vec{n}_{ijk}\| \, \|\vec{n}_{jkl}\|}. \tag{1.46}$$

The dihedral angles of a polypeptide main chain are conventionally referred as

$$\phi_i := \angle(C'_{i-1} - N_i - (C_\alpha)_i - C'_i), \quad \psi_i := \angle(N_i - (C_\alpha)_i - C'_i - N_{i+1}),$$
$$\text{and} \quad \omega_i := \angle((C_\alpha)_i - C'_i - N_{i+1} - (C_\alpha)_{i+1}),$$

where the index i denotes the residue number and sometimes can be omitted. The dihedral angles corresponding to the bonds of the side chains are labeled consequently as χ_i^1, χ_i^2 etc., starting from the bond to C_α (Fig. 1.21).

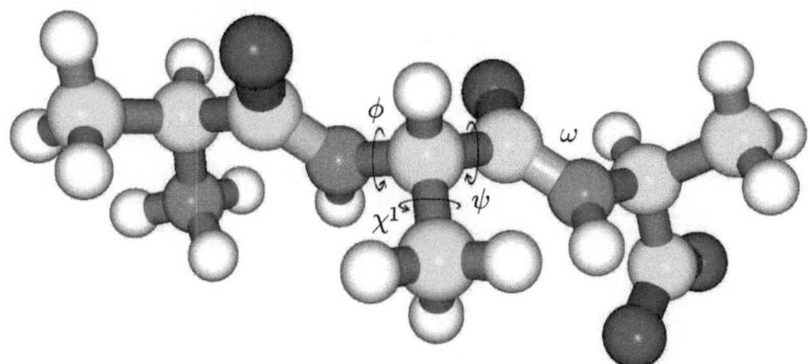

FIG. 1.21: Alanine tripeptide with peptide bonds colored in light green. Arrows at ϕ and ψ show the direction of rotation of the carboxyl end, which is on the right, with increment of dihedral angles. Values of side chain dihedral angles increase when the corresponding bonds are rotated clockwise, viewed from the main chain. ω is fixed at the value π in the most cases.

---

*Here and further in the work the sign function is defined as follows:

$$\text{sign}(x) := \begin{cases} 1, & \text{if } x \geq 0 \\ -1, & \text{if } x < 0. \end{cases} \tag{1.43}$$

Note that its value at zero is important for the correct determination of dihedral angles equal to π.

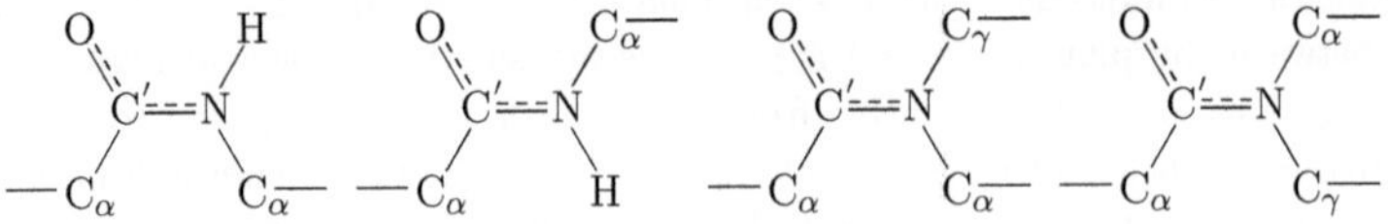

$$
\begin{array}{cccc}
\text{(a)} & \text{(b)} & \text{(c)} & \text{(d)}
\end{array}
$$

FIG. 1.22: Schematic representation of *cis* and *trans* conformations of peptide groups. The bonds of C_α and C_γ to the side chains and to hydrogen atoms are not shown. *Cis* (a) and *trans* (b) conformation of a peptide group when the second* involved residue is not proline. (c) and (d) show the same for a peptide group formed by any residue followed by proline.

Peptide bond has a partially double character owing to the resonance between two forms:

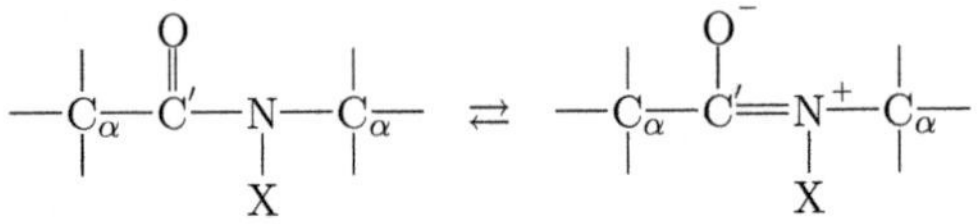

Here X stays for C_γ in proline and for H in any other residue. Due to delocalization of π-electrons, four atoms of the *peptide group* $(-C(=O)NH-)$ and two α-carbons always stay in one plane, either in *cis* ($\omega = 0$) or *trans* ($\omega = \pi$) conformation (Fig. 1.22, 1.23).

The most of peptide groups in native proteins occur in the *trans* conformation. If the second residue involved in the peptide bond is not proline, the probability of the *cis* arrangement is only about 10^{-3}. Such peptide groups are found, for example, in ser 197 – tyr 198, pro 205 – tyr 206 and arg 272 – asn 273 of carboxypeptidase A [17].

By contrast, peptide groups involving proline in the second position appear in *cis* form in $0.05 - 0.3$ of cases [17, 30]. This effect is sometimes explained in literature by the fact that the *cis* form in Figure 1.22(a) is energetically disadvantageous because of prohibitive proximity of the α-carbons. However, if the second residue is proline, carbons come close in both conformations (see Figure 1.22(c, d)). This justification can be disputed, because the van der Waals energy difference can be more than compensated[†] by the electrostatic interaction between an oxygen and a hydrogen atom, which come close in the *cis* form depicted in Figure 1.22 (a).

An alternative explanation could be that the *trans* arrangement is forced in the reaction of peptide bond formation. *Cis-trans* isomerization is relatively slow compared to the folding time of a protein. Once the protein is folded, the isomerization is hindered, since in the most cases it would require disruption of a stable structure.

---

*The residues are considered in the order as they are appended in protein synthesis, i. e. starting from amino end.

†According to the computations performed on SiViProF for the structures depicted in Figure 1.22 (a, b) with the permittivity of the protein hydrophobic core and Gasteiger [26, 27] partial charges.

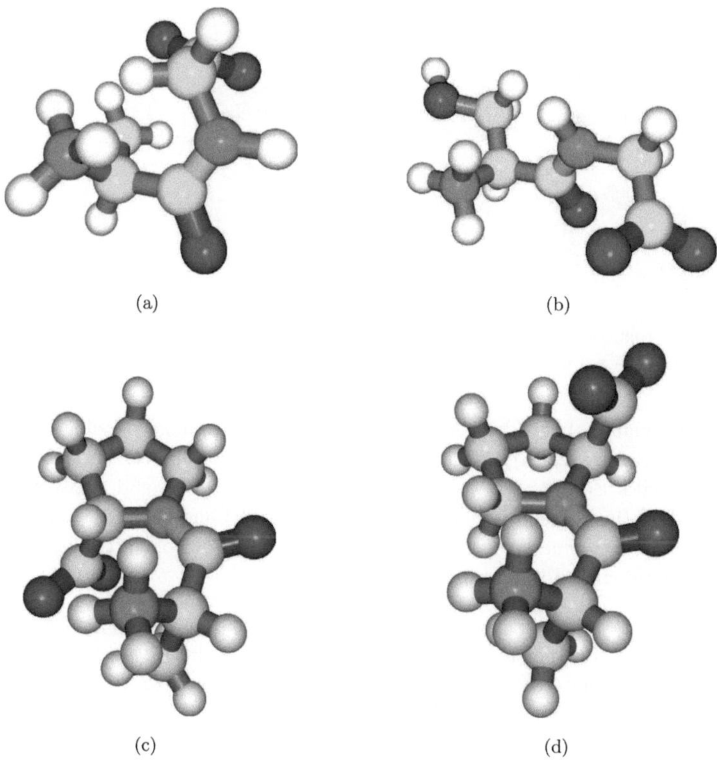

(a)

(b)

(c)

(d)

FIG. 1.23: *Cis* (a, c) and *trans* (b, d) conformations of peptide groups in serine-glycine (a, b) and alanine-proline (c, d) dipeptides.

The partial double bond character of $x-$pro peptide bonds, where x denotes any residue, is somewhat less expressed, since C_γ is more electronegative than a hydrogen atom. Therefore, the transition of the *trans* to the *cis* form in these cases may happen more often. Apart from that, there are enzymes that catalyze isomerization of $x-$pro peptide groups and facilitate folding [17].

The dihedral angles ϕ and ψ are relatively free to change. Therefore, the conformation of the protein backbone is mainly given by the values of these angles in each amino acid residue. However, there are some restrictions on twisting about the related bonds due to mutual repulsion of approaching atoms.

In a proline residue, the dihedral angle ϕ is fixed to certain values, and only ψ is variable (see Figure 1.24). In fact, we shall make a note that proline is also a special case in some other sense. Its ring is not aromatic and therefore not constrained to

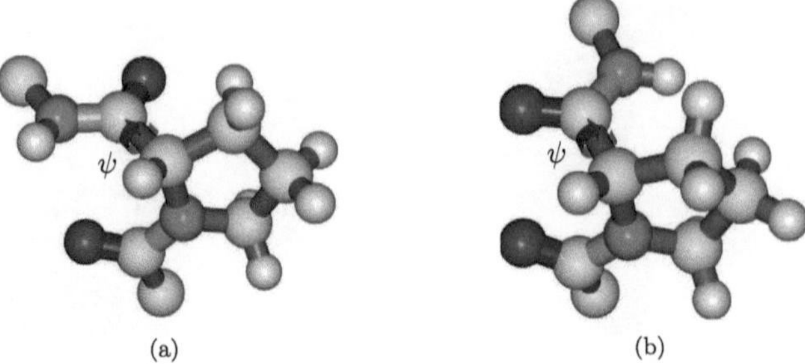

FIG. 1.24: Proline residues with flanking peptide groups. The only variable dihedral angle ψ is marked by a red arrow: (a) up puckering, (b) down puckering.

lay in one plane. Instead, there are two different puckerings possible. The *up puckering* (Figure 1.24(a)) is characterized by negative values of χ^1 and χ^3 (i.e. dihedral angles $\angle(N-C_\alpha-C_\beta-C_\gamma)$) and $\angle(C_\beta-C_\gamma-C_\delta-N)$ and positive values of χ^2 and χ^4 (i.e. dihedral angles $\angle(C_\alpha-C_\beta-C_\gamma-C_\delta)$ and $\angle(C_\gamma-C_\delta-N-C_\alpha)$), while for the *down puckering* (Figure 1.24(b)) the opposite holds. The both states are equally populated in proteins [30].

The puckering enforces corresponding values for ϕ angles in proline, or vice versa. According to a study of L. Vitagliano *et al.* [30], the average values of ϕ in proline residues encountered in native protein structures amount to $-58.7°$ for the up puckering and to $-69.8°$ for the down puckering.

For a peptide fragment containing only a proline residue with flanking naturally rigid peptide groups, as in Figure 1.24 (a,b), the dihedral angle ψ is the only degree of freedom, provided that the bond angles and lengths are fixed at their equilibrium values, and their high frequency fluctuations are neglected. If we let ψ to change from $-\pi$ to π and record the resulting van der Waals energy for this fragment, we obtain* a curve similar to those shown in Figure 1.25 (a,b). Peak values become lower, if we allow some rearrangements with minor changes of bond angles and lengths, or even flipping between the two puckerings.

The first peak arises when the hydrogen of a peptide group approaches a hydrogen at the β-carbon. This peak is present in the energy curves for the both puckerings, but in the up arrangement these atoms come closer. A neighboring small peak in the second energy curve corresponds to the interaction of the same amide hydrogen with a hydrogen at the γ-carbon, which is more remote in the up puckering. The last peak appears from the interaction of the amide hydrogen with the carboxyl carbon of the next peptide group. The strength of this interaction depends on the ϕ value in the central residue.

---

*Computations were performed by SIVIPROF with parameters listed in Appendix D.

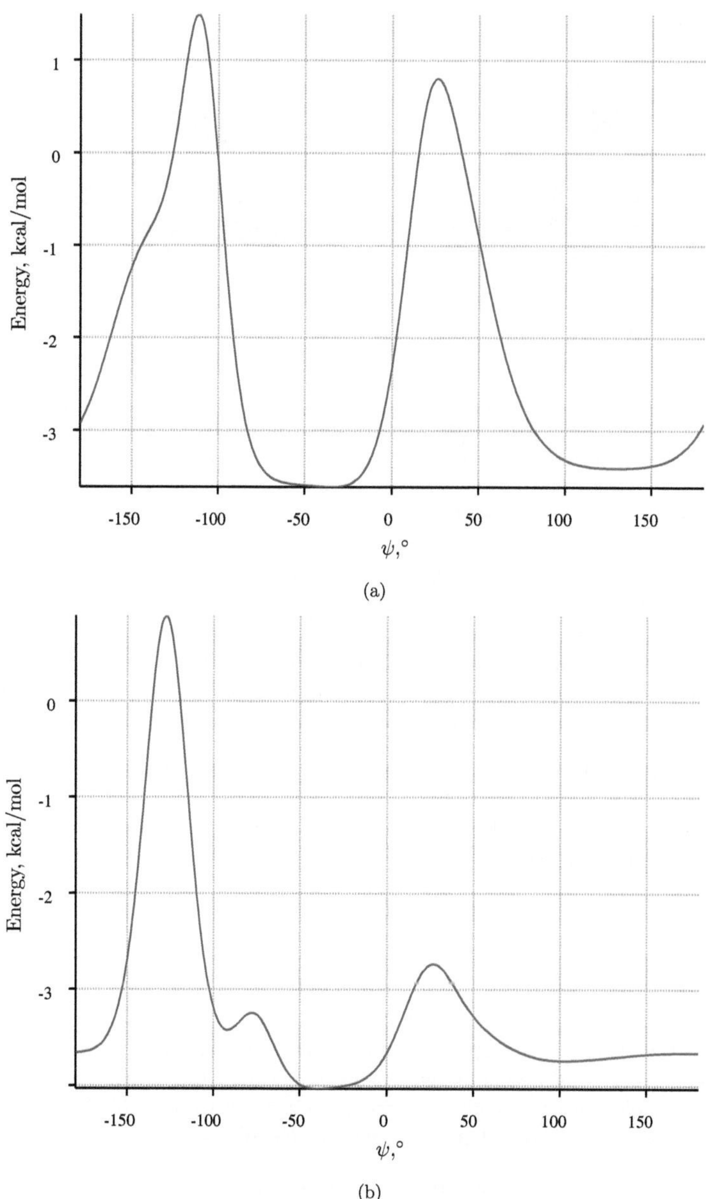

FIG. 1.25: van der Waals energy of the peptide fragments depicted in Figure 1.24: (a) with the up puckering, (b) with the down puckering.

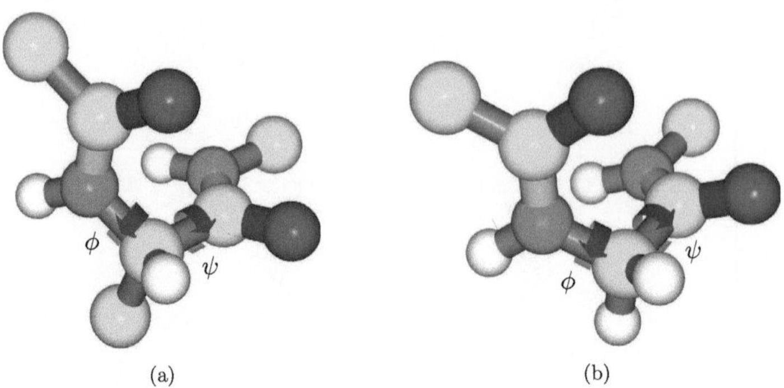

FIG. 1.26: Typical peptide fragments for exploration of the energy surface. Dihedral angles ϕ and ψ marked by red arrows can change. Other geometry features are considered to be essentially fixed at physiological conditions. (a) A peptide fragment typical for the most residues except glycine and proline. The side chain is reduced to the β-carbon. (b) The central fragment of a glycine-tripeptide.

In order to understand, to which extent the observed energy barriers restrict the twisting about the $C_\alpha C'$-bond (i.e. the variation of ψ) in a proline residue, we shall note that according to the course of lectures by A. V. Finkelstein and O. B. Ptitsyn [1], the typical range of thermal energy fluctuations is $k_B T$, or $\bar{R}T$ for a mole of substance, and any effect resulting in energy difference below this value can be disregarded. The $\bar{R}T$ value amounts to about 0.60 kcal/mol for $T = 300$ K.

The *equipartition theorem* of classical mechanics (see, for example, [14], p.31) states that the average value of each quadratic contribution to the energy (such as translational and rotational), associated with one degree of freedom, at a given temperature remains the same and equal to $\frac{1}{2}k_B T$ (or $\frac{1}{2}\bar{R}T$ for a mole of substance).

On the other hand, according to one of the formulations of the Arrhenius law, the mean frequency of transition over the energy barrier ΔE from one local minimum to another is given by:

$$\nu = \frac{k_B T}{h} e^{-\Delta E / k_B T}$$

where h is the Planck's constant (see [5] and references therein for more information about this relation and its limitations). This implies that the energy barrier $k_B T$ (or $\bar{R}T$ for molar energies) can be overcome in average within 4.35×10^{-13} s for $T = 300K$, while the barrier of $10\,k_B T$ – within 7.08×10^{-8} s.

These predictions appear to disagree with the statement of A. V. Finkelstein and O. B. Ptitsyn, saying that the typical difference of free energy in a native and a denaturated form of a protein is about 10 kcal/mol, and any effect resulting in larger energy difference can be considered as destructive (see [1], p. 58).

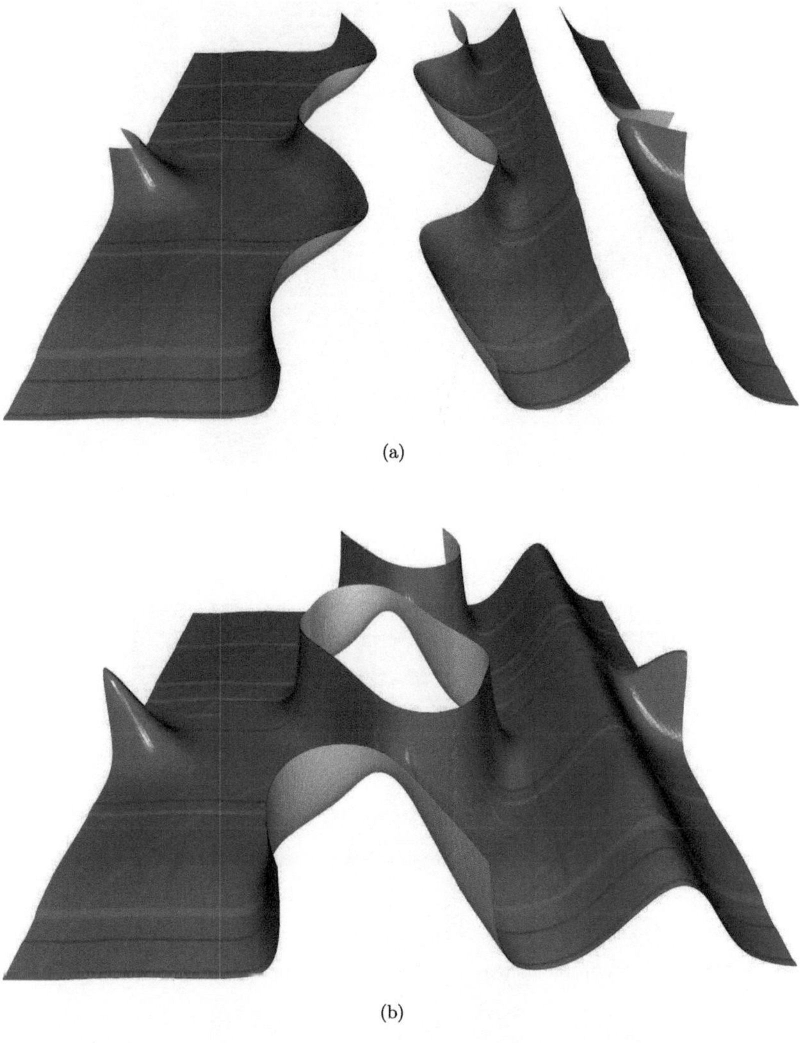

(a)

(b)

FIG. 1.27: van der Waals energy surface of the peptide fragment shown in Figure 1.26 (a):
(a) clipped at 5.96 kcal/mol ($10\,\bar{R}T$) above the global minimum; (b) clipped at 11.92
kcal/mol ($20\,\bar{R}T$) above the global minimum. ϕ value changes from $-180°$ at the left
to $180°$ at the right, ψ value changes from $-180°$ at the front to $180°$ at the back side. The
computed energy values in the depicted parts of the surface fall between -0.41 and 5.55
kcal/mol for (a), and between -0.41 and 11.51 kcal/mol for (b). Scratch-like irregularities
on the surface appear due to the cutoff equal to 4 Å and rounding errors.

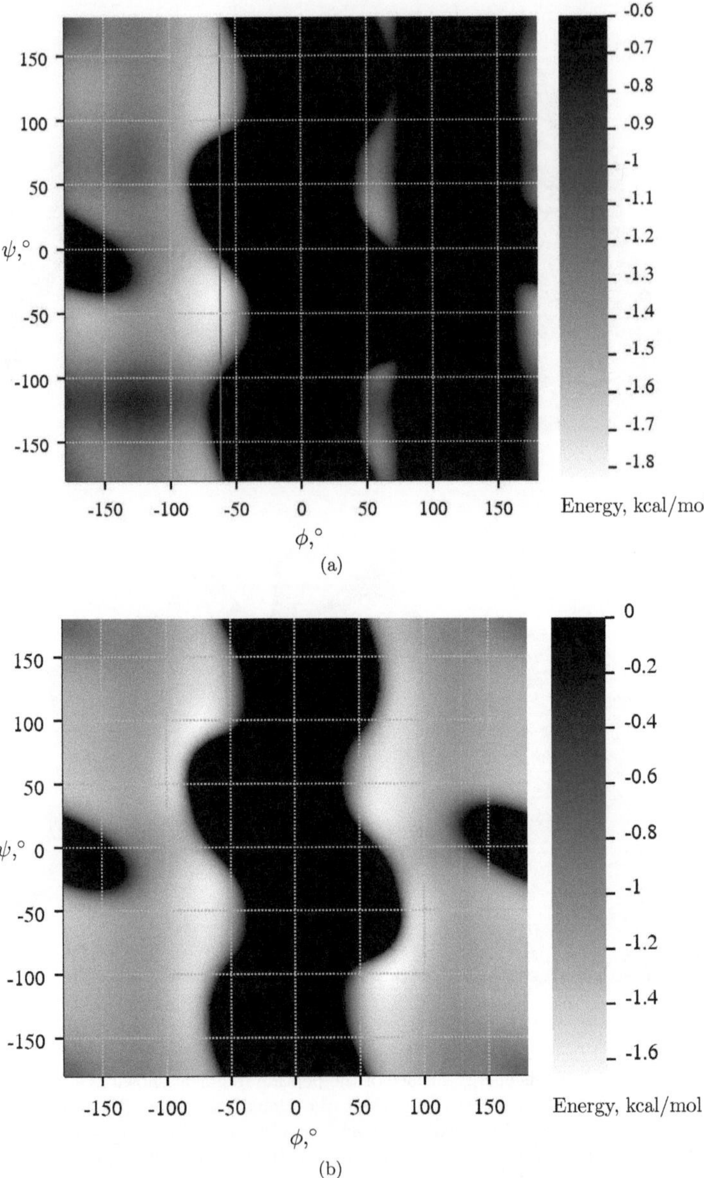

FIG. 1.28: van der Waals energy surfaces of peptide fragments, shown in Figure 1.26 (a) and (b) respectively. Pictures are generated by SiViProF with a specified cutoff for maximal energy value. Black color indicates high energy values. An approximate value of the dihedral angle ϕ in proline residues is marked in (a) by a red line.

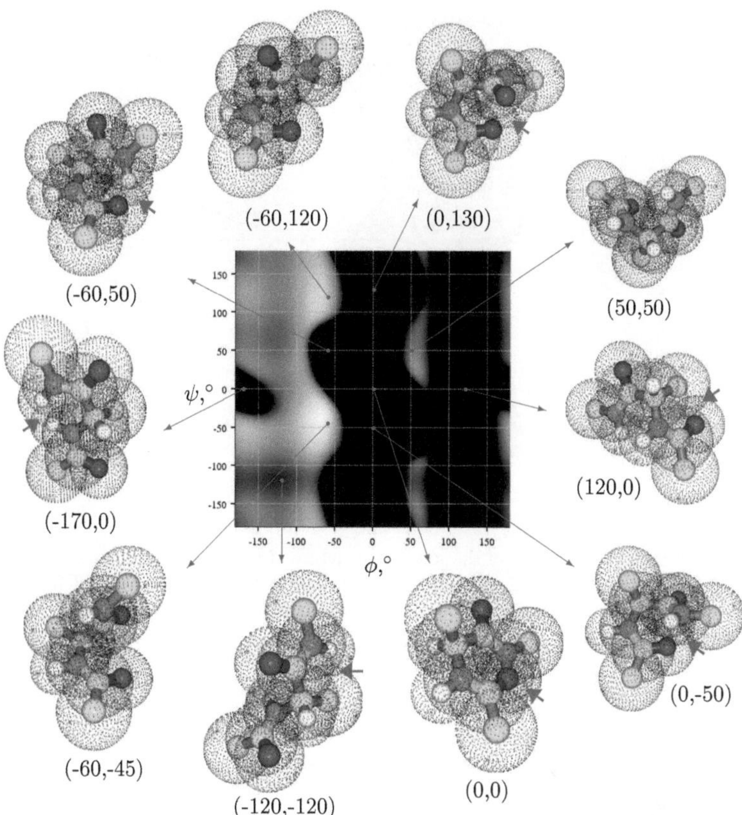

FIG. 1.29: Some conformations of a peptide fragment and the corresponding points on the van der Waals energy surface. Major distance violations are marked by short thick arrows.

In any case, it is known that transitions in proteins are cooperative, and the conformations of other residues in the chain can play crucial role in this process. In other words, implications that follow solely from consideration of the energy landscape of a separate residue are not sufficient to exclude that certain transition is likely to happen systematically within the folding time. Nevertheless, we shall see later that exploration of energy landscapes of typical protein fragments may be helpful for understanding why certain conformations of residues are more probable, while some others are hardly observed in native structures.

As mentioned before, the conformations of other residues depend on at least two dihedral angles, ϕ and ψ (see Figure 1.26). Besides, most of the non-proline residues have flexible side chains with additional degrees of freedom. The exception is glycine, the side chain of which consists of only a hydrogen atom.

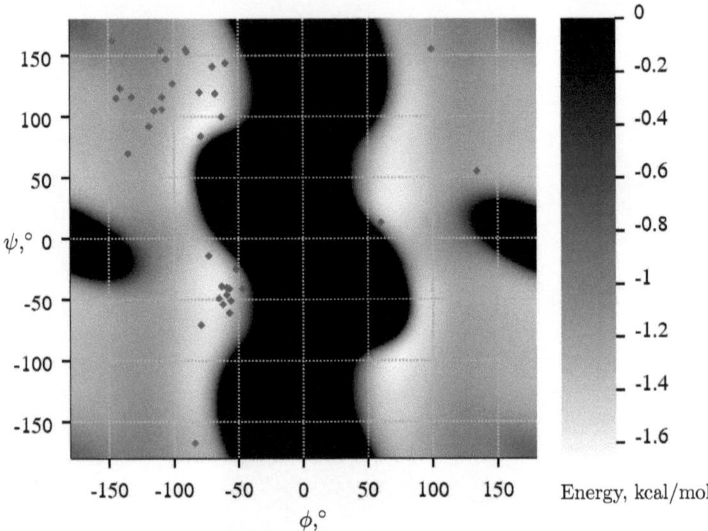

FIG. 1.30: Ramachandran map of the solution structure of the potassium channel inhibitor agitoxin 2 [31] with the van der Waals energy surface of the triglycine central fragment on the background. The conformation of each residue is marked by a red dot. Terminal residues are not shown.

If we let ϕ and ψ in the typical protein fragment, which is depicted in Figure 1.26 (a), to change from $-\pi$ to π and record the corresponding van der Waals energy, we obtain the surface shown in Figure 1.27. The same surface, also clipped at a certain energy value above the minimum, is shown as a color-coded projection in Figure 1.28 (a). A similar projection of the van der Waals energy surface obtained for a glycine residue with flanking peptide groups (Figure 1.26 (b)) is shown in Figure 1.28 (b). We see that certain residue conformations correspond to high energies, resulting from excessive proximity of some atoms. The region corresponding to positive values of ϕ is less accessible for non-glycine residues, because it corresponds to a clash of the β-carbon with the oxygen atom of the preceeding residue. Figure 1.29 gives some more detailed explanation for the appearance of peaks in the surface depicted in Figures 1.27 and 1.28 (a).

Inclusion of electrostatic energy does not change the picture significantly. In particular, it does not enable major violations of minimal separations, dictated by the van der Waals interaction, but introduces further restrictions for energetically favorable conformations. A more detailed discussion on this topic is given in Chapter 4.

If we consider a known protein structure and mark the conformation of each residue on the plot, we shall see that most points fall into favorable regions (see Figure 1.30). Such diagrams, mapping the protein structure to a set of points with coordinates

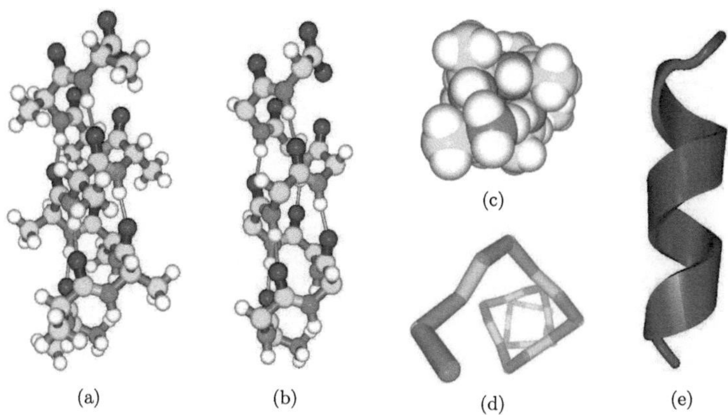

(a)   (b)   (c)   (d)   (e)

FIG. 1.31: The α-helical arrangement of deca-alanine. Thin light blue tubes denote hydrogen bonds. (a) The all-atom ball-and-stick representation. (b) The ball-and-stick model, where only the main chain atoms are shown. (c) A view from the amino end at the space-filling model. (d) The configuration of the main chain. (e) The ribbon model of the same structure.

(ϕ, ψ) that reflect the conformation of each residue, are called *Ramachandran plots*[*]. Typically they do not have an energy surface on the background, but rather a very rough sketch outlining preferable and forbidden regions (see [1]).

1.6.2 ELEMENTS OF SECONDARY STRUCTURE

As already mentioned in Section 1.2, the primary structure of a protein is given by the amino acid sequences of its constituting chains. When a polypeptide chain starts folding, clusters of amino acid residues organize themselves into regular structures, stabilized by their characteristic hydrogen bonding patterns (see Figures 1.31-1.34). Such elements adopt a certain main chain conformation, which determines the local *secondary structure* of the chain.

There are two major types of a regular secondary structure, which are abundant in proteins: α-helices (Figure 1.31) and β-sheets (Figure 1.32). In an *α-helix*, the oxygen of the i-th peptide group forms a hydrogen bond with the hydrogen of the $(i + 3)$-th peptide group (see Figure 1.31 (a, b)). In other words, a hydrogen bond is formed between i-th and $(i + 4)$-th residue, since the hydrogen atom of the same peptide group belongs to the next residue. The thereby created loop contains 13 atoms. Therefore, according to the customary nomenclature, an α-helix is a 4_{13}-helix. Native α-helices are right-handed, i.e. they approach the viewer by twisting counterclockwise (see Figure 1.31 (d)). The dihedral angles ϕ and ψ, typical for this structure, are respectively about $-60°$ and $-45°$. Certain variations

---

[*]Also Ramachandran maps or Ramachandran diagrams.

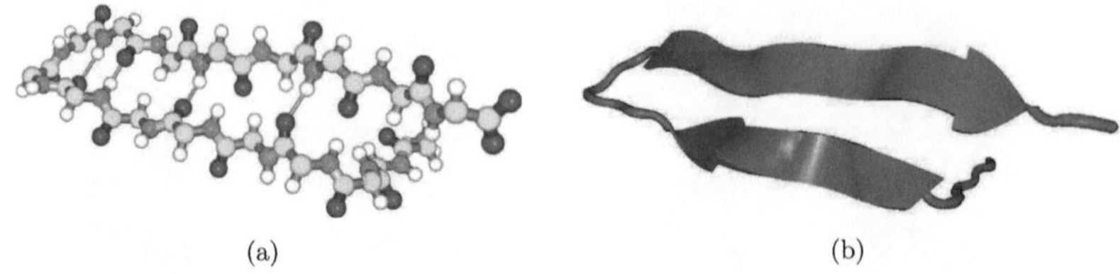

(a)   (b)

FIG. 1.32: The antiparallel β-sheet structure of the antimicrobial protein tachyplesin I, deduced by means of nuclear magnetic resonance spectroscopy. Atomic coordinates are obtained from RCSB Protein Data Bank, record 1WO1 by M. Mizuguchi *et al.* [32]. (a) A ball-and-stick model, where the only main chain atoms are shown. (b) The ribbon model.

of these values are possible. The most preferable of them are such that the sum of ϕ and ψ is approximately $-105°$. Points that correspond to such conformations on a Ramachandran map fall into the most favorable region, even if the energy of hydrogen bonding is not counted (see Figure 1.33). We shall note that the background energy surface in Figure 1.33 is computed for a fragment consisting of one isolated residue with the flanking peptide groups. The results of similar computations performed with SIVIPROF (see Chapter 4) for a periodic structure and for fragments including more residues show that the character of the landscape does not change, only the minima may become significantly deeper due to hydrogen bonding.

A left-handed 4_{13} helix is also possible, for example, with ϕ and ψ values around $60°$ and $45°$ respectively. However, this kind of helix is not encountered in native proteins. As far as one can judge based on Figure 1.33, this conformation is not so readily achievable.

Other types of helices exist, but most of them only hypothetically. They are discussed, for example, in [1] and [33]. In native folds one can find short fragments of right-handed 3_{10}-helices, which fall near the region of an α-helix on a Ramachandran plot (see Table 1.8 and Figure 1.33), but close to an edge of the favorable region.

TABLE 1.8: Regular secondary structures typical for native folds*.

Structure	Hydrogen bonding	ϕ	ψ
right-handed α-helix (α_R)	$(-C-O)_i \cdots (H-N-)_{i+4}$	$-60°$	$-45°$
right-handed 3_{10}-helix (($3_{10})_R$)	$(-C-O)_i \cdots (H-N-)_{i+3}$	$-50°$	$-25°$
antiparallel β-sheet ($\beta^{\uparrow\downarrow}$)	between antiparallel strands	$-135°$	$150°$
parallel β-sheet ($\beta^{\uparrow\uparrow}$)	between parallel strands	$-120°$	$135°$
poly(pro)II-helix	—	$-80°$	$155°$

---

*The values of ϕ and ψ are taken from [1].

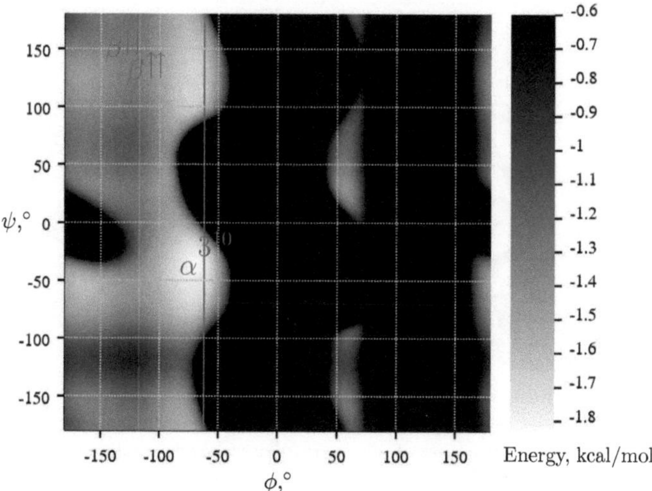

FIG. 1.33: Conformations typical for different secondary structure fragments with the van der Waals energy surface of a tripeptide central fragment including β-carbon (Fig. 1.26(a)). An approximate value of the dihedral angle ϕ in proline residues is marked by a red line.

The value of ϕ in proline residues is appropriate for their incorporation into an α-helix. However, on the place of the amide hydrogen, which would form a hydrogen bond with the nearby oxygen and stabilize the helix, it has a heavy carbon, giving raise to a strong steric repulsion of the next helical turn. Therefore, proline residues are destabilizing elements for an α-helical structure and can rather be found at its N-terminal [1]. Nevertheless, proline residues are capable to form regular arrangements without hydrogen bonding, such as poly(pro)I-helix (with *cis* conformation of peptide groups) and poly(pro)II-helix, realized in collagen.

In a *β-sheet*, hydrogen bonds are formed between different strands of an extended chain (Figure 1.32). β-sheets can be roughly subdivided into *parallel* and *antiparallel*, depending on whether the direction of neighboring strands is rather the same or the opposite. Mixed β-sheets are also possible. Examples of antiparallel and parallel β-sheets are visualized in Figure 1.34, (a) and (b) respectively.

Different side chains have unequal propensities to be found either in an α-helix or in a β-sheet. For example, polyalanine is known to readily form an α-helical structure. Tyrosine has an expressed tendency to be a part of a β-sheet and is unlikely to be found in a helix. Proline and negatively charged aspartic and glutamic acid residues are often located at the N-terminal of a helix, while positively charged histidine, lysine and arginine residues prefer helical C-terminals (see [1, 36], as well as Table 2.1 in the next chapter). Nevertheless, interactions with other residues and with the surrounding solvent are usually crucial for the final conformation.

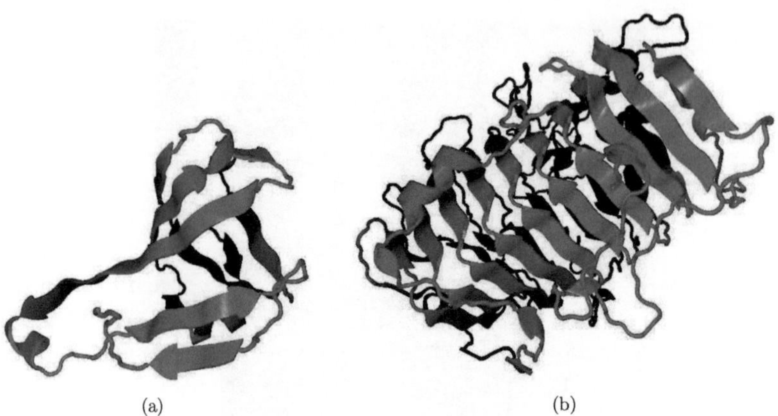

(a)   (b)

FIG. 1.34: The crystal structures of an aspartate decarboxylase subunit with antiparallel β-sheets (a) and of pectate lyase with parallel β-sheets (b), visualized in the form of ribbon models. Atomic coordinates are obtained from RCSB Protein Data Bank, records 1UHE by B. E. Lee *et al.* [34] and 3KRG by A. Seyedarabi *et al.* [35] respectively.

Irregular secondary structure is in general referred as a *coil*. However, one can segregate short standard fragments, such as β-turns or β-bulges.

1.6.3   TERTIARY AND QUATERNARY STRUCTURE OF PROTEINS

The *tertiary structure* of a protein is the final spatial arrangement of its folded chain. The tertiary structure is given by the atomic coordinates. One or another type of the secondary structure may prevail in the final fold, but usually it is a mixture of α-helices and β-sheets, intervened by coil fragments.

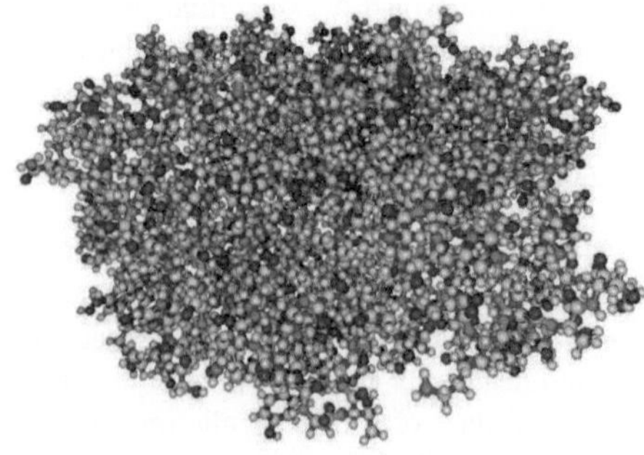

FIG. 1.35: The ball-and-stick model of the structure shown in Figure 1.34 (b), slightly rotated about the view direction.

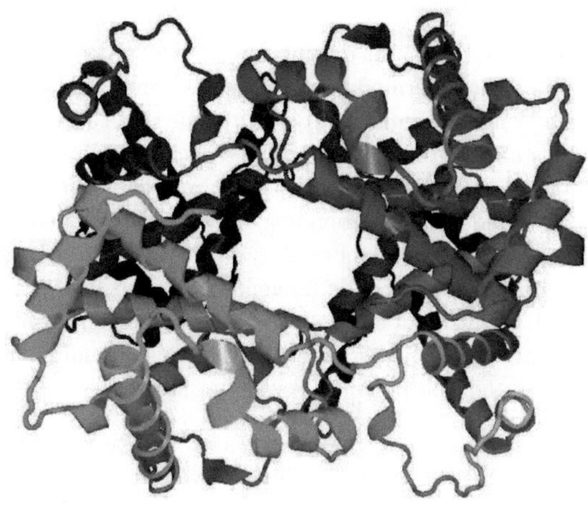

FIG. 1.36: The quaternary structure of human deoxyhaemoglobin, visualized in the form of a ribbon model. Atomic coordinates are obtained from RCSB Protein Data Bank, record 2HHB by G. Fermi *et al.* [37].

Analysis of the protein structure is greatly facilitated by *ribbon models*, in which α-helices are represented by winding ribbons (see Figure 1.31 (e)), and β-strands are depicted using ribbons in the form of arrows showing the direction of the chain growth (see Figure 1.32 (b)). It can be well understood, if one compares Figure 1.34 (b) with Figure 1.35, depicting the same structure.

The quaternary structure is the organization of several chains or subunits into a functional complex. For example, haemoglobin is a heterotetramer consisting of two α and two β subunits*, which are hold together by salt bridges, hydrogen bonds and hydrophobic interactions (see Figure 1.36).

1.7 INFLUENCE OF ENVIRONMENT

Since the folding of a protein chain occurs not in vacuum, external factors play, beside intramolecular interactions, a crucial role in determination of the final structure. One of such factors is the interaction with the surrounding solvent. As discussed in Section 1.2, amino acid residues have different properties with respect to their behavior in an aqueous environment. Some of them are capable to form hydrogen bonds with water molecules or even ionize, others have relatively non-polar hydrocarbon side chains with properties resembling those of petrol. Peptide fragments consisting of such residues are poorly mixable with water and have a tendency to collect together, giving raise to the *hydrophobic effect*.

---

*Here α and β are customary subunit names, not related to their secondary structure.

1.7.1 THE HYDROPHOBIC EFFECT

The hydrophobic effect is the major driving force in the process of protein folding. One can show by means of thermodynamical experiments, that at the room temperature its nature is mainly entropic [1]. We shall recall from Subsection 1.5.5 that hydrogen bonds have significant energy, and each water molecule is capable of forming up two four hydrogen bonds. Non-polar molecules disrupt the very flexible but almost saturated hydrogen-bonding network of water and thereby force its rearrangement, such that the number of hydrogen bonds can be still preserved. An ice-like layer with hydrogen bonds tangential to the surface of the non-polar compound is formed, resulting in entropy loss. Free energy increases with decreasing entropy (see equation (1.7)), therefore the adjustments minimizing the contact surface between water and non-polar molecules are favored.

In principle, the solubility of side chains in water can be measured in partition experiments that explore the equilibrium distribution of related chemical species* in two or three phases, one of which is a non-polar liquid and another one is water. The free energy of transfer of one mole of a compound from a non-polar solvent to water is given by

$$\Delta G^{[n \to w]} = -RT \ln \left(\frac{c^{[w]}}{c^{[n]}} \right), \tag{1.47}$$

where $c^{[n]}$ and $c^{[w]}$ are respectively the concentrations of the dissolved substance in the two mentioned phases. However, it is not easy to perform such experiments [1], because purely non-polar compounds hardly dissolve in water, while charged molecules avoid the hydrophobic phase. Therefore, in place of one or another phase, an amphiphilic substance is typically used. Presumably, this yields transfer energies with too small absolute values, particularly for ionizable amino acids. Besides, the results obtained with different solvents show substantial discrepancy.

As stated in a review by K. M. Biswas *et al.* [38], the first major amino acid hydrophobicity scale, which became one of the most cited, was proposed by Y. Nozaki and C. Tanford [39]. They used ethanol and dioxane to imitate the protein interior. However, the set of the amino acids, included in this study, was not complete. J.-L. Fauchère and V. Pliška were among the first researchers who used amino acid derivatives and introduced a hydrophobicity scale for the full amino acid set. Amino acids were represented by their N-acetyl amides. In place of a hydrophobic solvent octanol was used, which has a relatively long hydrocarbon chain and a hydroxyl group capable of hydrogen bonding. The transfer energies given by this scale, as reproduced in [40], are summarized in the second column of Table 1.9.

A number of other scales was proposed since then. E. Q. Lawson *et al.* [41] suggested that N-cyclohexyl-2-pyrrolidone is a better candidate to mimic the protein core, as

---

*Generally, instead of native amino acids, always containing ionizable amino and carboxyl groups, chemicals representing only the residues, or their side chains, are used.

TABLE 1.9: Free energies of amino acid side chain transfer from apolar solvent to water.

Residue	$\Delta G^{[C_8H_{17}OH \rightarrow H_2O]}$, kcal/mol*	$\Delta G^{[chp \rightarrow H_2O]}$, kcal/mol†
alanine	0.42	-0.48 ± 0.06
arginine	-1.37	-0.06 ± 0.08
asparagine	-0.82	-0.87 ± 0.06
aspartic acid	-1.05	-0.75 ± 0.09
cysteine	$1.34^\ddagger$ (2.10)	-0.32 ± 0.07
glutamic acid	-0.87	-0.71 ± 0.07
glutamine	-0.30	-0.32 ± 0.06
glycine	0.00	0.00
histidine	0.18	-0.51 ± 0.06
isoleucine	2.46	0.81 ± 0.07
leucine	2.32	1.02 ± 0.09
lysine	-1.35	-0.09 ± 0.06
methionine	1.68	0.81 ± 0.06
phenylalanine	2.44	1.03 ± 0.06
proline	0.98	2.03 ± 0.06
serine	-0.05	0.05 ± 0.04
threonine	0.35	-0.35 ± 0.07
tryptophan	3.07	0.66 ± 0.06
tyrosine	1.31	1.24 ± 0.08
valine	1.66	0.56 ± 0.08

it contains a fragment resembling a peptide group. Besides, the physical properties of this solvent, such as the dielectric constant, viscosity, and a few others, are similar to those of a protein interior. The experiments involved amino acids in their free and hydrochloride forms. The values obtained in this study are listed in the third column of Table 1.9. The transfer energy of glycine is subtracted from all values, in order to elucidate only the contributions of side chains.

According to common observations, not only related to the hydration energies of the amino acid side chains, the hydrophobic effect increases roughly proportionally to the surface area of non-polar compounds. As proposed by B. Lee and F. M. Richards

---

*Determined by J.-L. Fauchère and V. Pliška free energies of transfer of amino acid side chains from octanol to water, as reproduced in [40].

†Free energies of side chain transfer from N-cyclohexyl-2-pyrrolidone to water, obtained by E. Q. Lawson *et al.* [41].

‡According to [17], this value is related to a half of cystine. The value for cysteine is given in brackets.

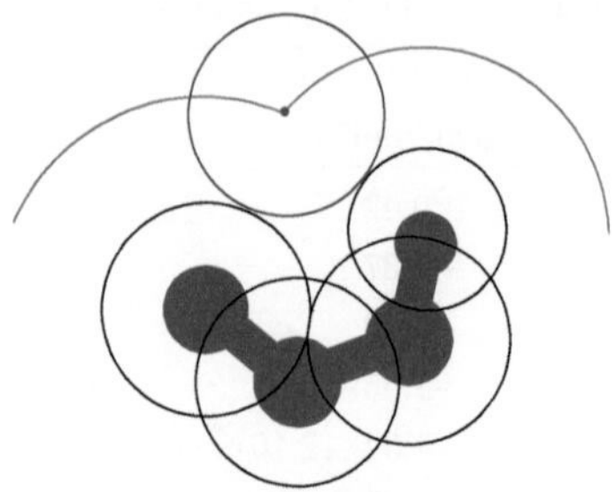

Fig. 1.37: Determination of solvent accessible surface (red). Black circles outline the van der Waals spheres of atoms. Blue circle denotes a probe sphere.

[42], and adopted by many other researchers, a solvent accessible surface is defined as the locus of the center of a probe sphere of the radius of $1.4\,\text{Å}$, approximating a water molecule, rolled over the van der Waals envelope of the solute (Fig. 1.37). The area of this locus is referred as the *solvent-accessible surface area* (SASA). The hydrophobic free energy is estimated to be approximately between 0.02 and 0.025 kcal per mole and Å^2 of non-polar SASA [1, 40].

1.7.2   Interaction of charged groups with solvent

Since water molecules are electric dipoles, they line up along charged surfaces, so that the entropy may decrease even more than in cases of nonpolar solutes. However, this generally does not result in aggregation of dissolved objects, as it happens with hydrophobic species. The reason for that is, apart from repulsion forces between univalent charges, a lower free energy of charges[*] in water surrounding.

A rough explanation of this effect can be given using classical electrostatics (see also [1] and [17]). The electrical free energy of a sphere with the radius R and charge q, suspended in a medium with the relative permittivity ϵ_r, is:

$$G_s = \frac{q^2}{8\pi\epsilon_0\epsilon_r R}.\tag{1.48}$$

Therefore, to transfer this charge from a solvent-exposed surface to a protein core, one has to spend the energy

$$\Delta G_s^{[s\to p]} = \frac{q^2}{8\pi\epsilon_0 R}\left(\frac{1}{\epsilon_p} - \frac{1}{\epsilon_s}\right),\tag{1.49}$$

where ϵ_p and ϵ_s are the relative permittivities of the medium in the protein core and at the surface respectively. If we take $R = 1.52\,\text{Å}$, which is the typically used

---

[*]We assume that the considered charges are sufficiently large to provide at least local water polarization.

van der Waals radius of oxygen (see Table D.1 in Appendix D), $\epsilon_p = 3$ and $\epsilon_s = 40$ (see Subsection 1.5.3 or [1]), we obtain $\Delta G_s^{[s \to p]} \approx 5.6 \times 10^{-23}$ kcal for a unit charge, or 33.7 kcal for one mole of such charges. This is a significant energy, compared to contributions of other effects. However, the transfer energies rapidly diminish with decreasing charge magnitude. For a partial charge about $0.2\,|e^-|$ with the same radius we obtain $\Delta G_s^{[s \to p]} \approx 2.2 \times 10^{-24}$ kcal, or 1.3 kcal/mol.

The following speculation suggests that even a transfer of a pair of two opposite charges with absolute values q_+ and q_- into a protein core is not favorable, although they interact more intensively in a hydrophobic environment. Their interaction energy is given by

$$G_{\pm}^{[s]} = -\frac{q_+ q_-}{4\pi\epsilon_0 \epsilon_s r}$$

at a water-protein interface, and by

$$G_{\pm}^{[p]} = -\frac{q_+ q_-}{4\pi\epsilon_0 \epsilon_p r}$$

in a protein core, implying the difference

$$\Delta G_{\pm}^{[s \to p]} = -\frac{q_+ q_-}{4\pi\epsilon_0 r}\left(\frac{1}{\epsilon_p} - \frac{1}{\epsilon_s}\right).$$

Even in a case of immediate proximity of charges, i.e. when $r = R_+ + R_-$, where R_+ and R_- are the van der Waals radii of the spherical charges q_+ and q_-, the solvation energy difference $\Delta G_s^{[s \to p]}$ exceeds the absolute value of $\Delta G_{\pm}^{[s \to p]}$ by more then 100%:

$$\frac{|\Delta G_{\pm}^{[s \to p]}|}{\Delta G_s^{[s \to p]}} = \frac{\frac{q_+ q_-}{4\pi\epsilon_0 r}\left(\frac{1}{\epsilon_p} - \frac{1}{\epsilon_s}\right)}{\frac{1}{8\pi\epsilon_0}\left(\frac{q_+^2}{R_+} + \frac{q_-^2}{R_-}\right)\left(\frac{1}{\epsilon_p} - \frac{1}{\epsilon_s}\right)} = \frac{\frac{q_+ q_-}{R_+ + R_-}}{\frac{1}{2}\left(\frac{q_+^2}{R_+} + \frac{q_-^2}{R_-}\right)} \leq \frac{1}{2}. \tag{1.50}$$

The last inequality in 1.50 is proven in the described below Proposition 1.2, which utilizes the following lemma:

LEMMA 1.1
For any $x, y \in \mathbb{R}$ $(x^2 + y^2 \neq 0)$ holds:

$$\frac{xy}{x^2 + y^2} \leq \frac{1}{2}. \tag{1.51}$$

Proof. Multiplying both sides of (1.51) by $2(x^2 + y^2)$ and collecting all terms yields:

$$(x - y)^2 \geq 0.$$

The latter holds $\forall\, x, y \in \mathbb{R}$. $\square$

PROPOSITION 1.2
For any $R_+, R_- \in \mathbb{R}^+ \setminus \{0\}$ and for any $q_+, q_- \in \mathbb{R}$ the following inequality is valid:

$$\frac{\frac{q_+ q_-}{R_+ + R_-}}{\left(\frac{q_+^2}{R_+} + \frac{q_-^2}{R_-}\right)} \leq \frac{1}{4}.$$

(1.52)

Proof. The left-hand side of 1.52 can be rewritten as:

$$\frac{\frac{q_+ q_-}{\sqrt{R_+ R_-}} \frac{\sqrt{R_+ R_-}}{R_+ + R_-}}{\left(\frac{q_+^2}{R_+} + \frac{q_-^2}{R_-}\right)}.$$

We note that both

$$\frac{\frac{q_+ q_-}{\sqrt{R_+ R_-}}}{\left(\frac{q_+^2}{R_+} + \frac{q_-^2}{R_-}\right)}$$

(1.53)

and

$$\frac{\sqrt{R_+ R_-}}{R_+ + R_-}$$

(1.54)

have the form of the left-hand side in (1.51), where $x = \frac{q_+}{\sqrt{R_+}}$ and $y = \frac{q_-}{\sqrt{R_-}}$, or $x = \sqrt{R_+}$ and $y = \sqrt{R_-}$ respectively. Applying Lemma 1.1 to expressions (1.53) and (1.54) immediately yields (1.52). $\square$

Since

$$|\Delta G_\pm^{[s \to p]}| = |G_\pm^{[p]}| \left(1 - \frac{\epsilon_p}{\epsilon_s}\right),$$

from 1.50 follows that

$$|G_\pm^{[p]}| \leq \frac{\Delta G_s^{[s \to p]}}{2\left(1 - \frac{\epsilon_p}{\epsilon_s}\right)}.$$

In particular, $|G_\pm^{[p]}| < \Delta G_s^{[s \to p]}$ for $\epsilon_p < 2\epsilon_s$. This treatment implies that also for a case when charges virtually do not interact at the protein surface and come close only in a protein core, their solvation is more favorable.

As a result, when a protein chain starts folding in an aqueous environment, it typically collapses into a globule, hiding hydrophobic residues and leaving ionizable and highly polar groups at the surface. Ionized groups are practically absent in a protein core [1]. A glance at Table 1.4 makes clear that it is more advantageous to uncharge an ionizable group before burying it into the protein interior, if such a transfer is necessary at all. This destabilizing process is generally avoided, with the exception of cases when it is necessary for protein functions.

1.7.3 CHAPERONES AND ASSISTED PROTEIN FOLDING

Although the validity of the Anfinsen's dogma (see Section 1.1) is often accepted in a general sense without specification of folding conditions, the former has some important limitations. First of all, there are evidences that protein folding *in vivo**

---

*In a living organism, literally, *in life* (Latin).

often occurs cotranslationally [43–45], which implies that sequential folding starting from the amino end, as well as a specific initial conformation, may be necessary for correct folding of some proteins. Refolding *in vitro* is often significantly more slow compared to intracellular folding, if possible at all [46–49]. Besides, the conformations of denaturated proteins that are capable to refold do not have to be completely random. A protein is considered to be denaturated if its structure is sufficiently different from its native form to impede its functions. Fragments of secondary or tertiary structure that are essential for correct folding may be preserved.

Second, certain factors, such as higher temperature, changed pH, or presence of some other molecules, may result in misfolding or denaturation of proteins. When a cell is subjected to such kind of stress, it responds by expressing increased amounts of *heat shock proteins*, which identify incorrectly folded proteins and mediate their refolding.

Heat shock proteins are a subgroup of molecular *chaperones*, which are proteins assisting in folding and assembly of some macromolecules or molecular complexes. Most chaperones are not substrate-specific and do not convey any steric information, although exceptions from this rule exist [50]. Some molecular chaperones bind to hydrophobic fragments on a protein surface and protect them from aggregation with other molecules. Another group, *chaperonins*, is characterized by a double-ring structure with a central cage inside. Their function is to capture misfolded proteins, stretch them and provide an environment for refolding [51].

Although many proteins are able to fold spontaneously, a significant number requires presence of chaperones. Apart from that, some enzymes are involved in protein folding, such as *protein disulfide isomerase*, assisting in formation of disulfide bonds, and *peptidyl-prolyl cis-trans isomerase*, accelerating isomerization of $x-$pro peptide groups.

1.7.4 PRIONS

Prions are proteins that, in addition to their native cellular structure PrP^C, possess at least one more stable conformation, called *scrapie** form, PrP^{Sc}, which is characterized by capability to promote refolding of PrP^C into the PrP^{Sc} form.

The conversion to PrP^{Sc} involves a drastic change from an α-helical structure to an arrangement with a high β-sheet content [52, 53]. There are evidences that this transformation can be facilitated by chaperones in the presence of PrP^{Sc}. The scrapie molecules are less solvable and tend to aggregate. Their amount grows exponentially, causing an accumulation of the unsolvable mass and cell destruction. Diseases caused by prions include, among others, bovine spongiform encephalopathy (BSE), known as mad cow disease, scrapie in sheep, and Creutzfeldt-Jakob disease (CJD) in humans. All of them affect neural tissues and have a fatal outcome.

---

*Named after scrapie, a transmissible degenerative neural disease occurring among sheep and goats. One of its symptoms is compulsive scraping of infected animals against surrounding objects.

The scrapie isoform is remarkably stable. It is only partially hydrolyzed by proteases, whereas PrP^C is completely degraded under the same conditions [53]. It has been shown that CJD scrapie agents are exceptionally resistant to thermal treatment, with no significant changes after one hour of exposure to the temperature $132°$ C (see [53] and references therein). These facts suggest that the popular assumption relating the native structure to the global energy minimum, achievable from any random initial configuration, can be disputed.

1.8 PROTEIN SYNTHESIS

The information about the amino acid sequences of proteins that can be synthesized by a cell is encrypted in the genetic code by means of nucleotide triplets, called *codons*. Each codon corresponds to at most one amino acid, but the majority of the amino acids can be coded by a few different triplets (see Table 1.10). Some nucleotide triplets have no relation to any amino acid. They are *stop codons*, signalizing the end of the coding sequence and resulting in release of the synthesized protein.

TABLE 1.10: Rules for decoding genetic messages [54]. Capital letters denote conventional abbreviations for nucleotide names: A – adenine, C – cytosine, G – guanine, U – uracil*. Lower case abbreviations stand for three-letter codes of amino acids[†], specified in Table A.2.

		Second letter of codon			
		U	C	A	G
First letter of codon	U	UUU phe UUC phe UUA leu UUG leu	UCU ser UCC ser UCA ser UCG ser	UAU tyr UAC tyr UAA **stop** UAG **stop**	UGU cys UGC cys UGA **stop** UGG trp
	C	CUU leu CUC leu CUA leu CUG leu	CCU pro CCC pro CCA pro CCG pro	CAU his CAC his CAA gln CAG gln	CGU arg CGC arg CGA arg CGG arg
	A	AUU ile AUC ile AUA ile **AUG** met	ACU thr ACC thr ACA thr ACG thr	AAU asn AAC asn AAA lys AAG lys	AGU ser AGC ser AGA arg AGG arg
	G	GUU val GUC val GUA val GUG val	GCU ala GCC ala GCA ala GCG ala	GAU asp GAC asp GAA glu GAG glu	GGU gly GGC gly GGA gly GGG gly

*In DNA uracil is replaced by thymine (T).

[†]Except "**stop**" denoting stop codons.

The genetic material of a cell is stored in the form of deoxyribonucleic acids (DNA).[*] In *prokaryotic* cells, such as bacteria and archaea, DNA is usually suspended in cytosol in complex with regulatory proteins, forming irregular *nucleoid*[†]. In *eukaryotic* cells DNA is closely packed in chromatin together with proteins and ribonucleic acids (RNA), and surrounded by a double membrane, forming a *nucleus*.

When a protein is to be synthesized, the part of the DNA, containing the coding nucleotide sequence, is copied, or *transcribed*, into a messenger RNA (mRNA). In eukaryotes, the newly synthesized mRNA is subjected to certain processing and transported to cytosol, where the genetic message is decoded.

1.8.1 RIBOSOMES

The translation of a genetic code into a polypeptide chain is performed on *ribosomes*, cellular organelles of a size about $200 - 300$ Å, which are composed of RNA and proteins. Each ribosome consists of a large and a small subunit (see Figure 1.38), which associate together to start protein synthesis and go apart when it finishes. Prokaryotic and eukaryotic ribosomes are functionally very similar, although the latter are somewhat larger. Their sedimentation coefficients[‡], customary used for characterization of ribosomes and their parts, amount to 70S and 80S respectively. The sedimentation coefficients of prokaryotic large and small subunits are 50S and 30S, whereas the corresponding eukaryotic values are 60S and 40S.

The function of the small subunit is to ensure correct reading of the genetic code. The large subunit catalyzes peptide bond formation, protein release, and hydrolysis of guanosine triphosphate (GTP), which provides energy for acceleration of conformational changes[§]. The translocation of a ribosome to the next codon is performed by two subunits in cooperation [55].

Another feature of the large subunit is a tunnel that spans from the catalytic site, *peptidyl transferase center* (PTC), near the interface between subunits and opens at the opposite side (see Figure 1.38(e, f)). The opening of the tunnel corresponds to the site where nascent β-galactosidase emerges from ribosomes [56, 57]. Moreover, it has been shown that eukaryotic translocons, which help to transfer nascent proteins into the lumen of the *endoplasmic reticulum* (ER), align with the tunnel exit [58, 59] (see Subsection 1.8.4 for details). A number of antibiotics can block the tunnel entrance, thereby prohibiting synthesis of some proteins [60]. Besides, there are evidences that the narrowest part of the tunnel may act as a gate causing sequence-specific nascent chain elongation arrest in respond to cellular needs [55, 61–63].

---

[*]The genome of the most viruses is stored in RNA.

[†]The nucleoid of phylum *Planctomycetes* of the domain bacteria is enclosed into a single or double membrane, resembling an eukaryotic nucleus.

[‡]The values of sedimentation coefficients, measured in Svedberg (S) units, do not add up when particles bind together, since sedimentation rates are influenced by the surface area of particles.

[§]The translation is an exergonic process, which can proceed spontaneously without GTP consumption, however, relatively slowly [43].

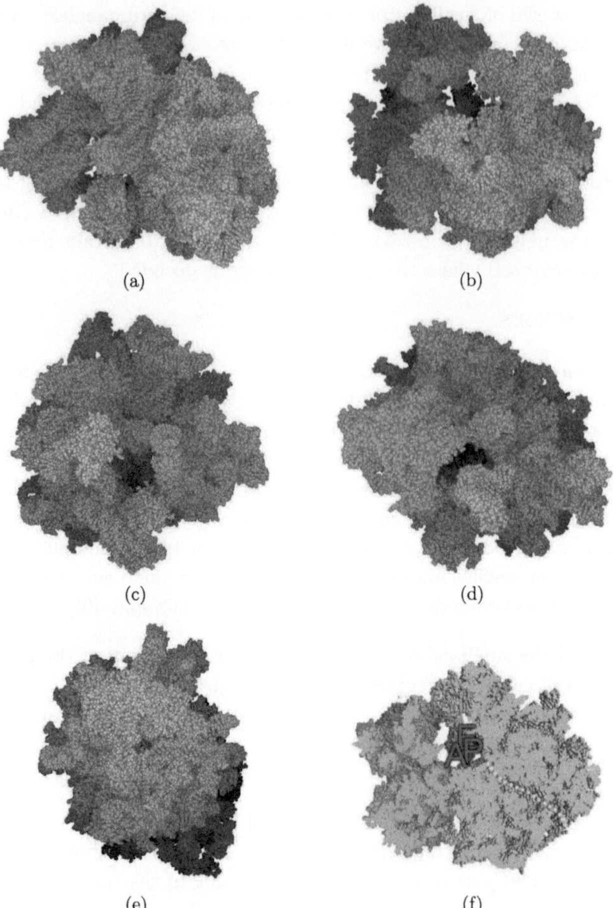

(a)   (b)

(c)   (d)

(e)   (f)

FIG. 1.38: The space-filling model of an eukaryotic ribosome from *Saccharomyces cerevisiae*. The atomic coordinates are deduced by C. M. Spahn *et al.* [64] through docking molecular models of RNA and protein components into a 11.7 Å cryo-EM[*] map. The structure is obtained from RCSB Protein Data Bank[†] records 1S1H and 1S1I, and visualized[‡] using SIVIPROF. The small 40S subunit is depicted in orange and the large 60S subunit is colored in light green. (a–d) Views from different sides along the interface between two subunits. (e) A view from the side of the large subunit with an opening of the protein exit tunnel in the center. (f) A cross-section through the ribosomal catalytic center and the protein exit tunnel[§]. Letters **A**, **P** and **E** denote respectively the locations of aminoacyl, peptidyl and exit sites. Yellow dot line marks the exit tunnel for nascent proteins.

---

[*]Cryo-EM is a form of electron microscopy working with frozen samples at low temperatures.

[†]See Section 1.10 for information about RCSB Protein Data Bank (PDB).

[‡]The PDB record does not contain positions of hydrogen atoms, therefore they are not shown.

[§]A view from the other side and upside down relative to the subfigure (a).

Most researches seem to share the view that all nascent chains pass through this tunnel. However, some experimental results appear not to support it. For example, synthesis of polyphenylalanine is not inhibited by erythromycin and streptogramin B, which block the ribosomal tunnel and thereby interrupt elongation of chains containing basic amino acids [60, 65]. L. A. Ryabova *et al.* [66] have shown that short amino terminal fragments of the phage coat protein, synthesized in absence of tyrosine, lysine, cysteine and methionine, can be accessible to antibodies near the groove along the interface between two subunits. A fluorescence study by W. D. Picking *et al.* [65] has indicated that polyphenylalanine and polylysine apparently do not enter the tunnel but rather escape to cytosol near the PTC*. It has been suggested that peptides may exit the ribosome by different routes, depending on amino acid markers at the amino end of the nascent chain [55].

1.8.2 MECHANISM OF TRANSLATION

The amino acids are delivered to the ribosome by transfer RNA (tRNA), relatively small RNA molecules folded similar to the one depicted in Figure 1.39. Each tRNA has a three-nucleotide region, called *anticodon*, responsible for pairing with mRNA codons. Another site of a tRNA molecule can be linked by an ester bond to a carboxyl group of a specific amino acid. This reaction is catalyzed by special enzymes, aminoacyl-tRNA synthetases, which take care that the tRNA with a certain anticodon is loaded, or *charged*, by the corresponding amino acid.

Ribosomes have three tRNA-binding sites, which are located on the interface between two subunits (see Figures 1.38(f) and 1.40): the *aminoacyl site* (**A** site), the *peptidyl site* (**P** site), and the *exit site* (**E** site). The small subunit directs the mRNA through a narrow groove to the decoding sites on the subunit contact interface, where mRNA codons can be matched through the complementary base pairing with tRNA anticodons.

The translation is initiated when the small subunit binds to mRNA in a way that the **P** site is occupied by the start codon. The role of the initiating codon is played by AUG, which is the only codon corresponding to methionine. In prokaryotes, the mRNA is positioned on the small subunit with the aid of the Shine-Dalgarno sequence, which is located upstream to the start codon. In this case, only a specific initiating tRNA, charged by N-formylmethionine, can bind to the start codon. In eukaryotes, the small subunit, accompanied by some protein factors, binds to a special tag at the 5'-end of mRNA and slides until the first AUG codon is found. The initiating tRNA from cytosol of eukaryotes is charged by methionine, although this tRNA is different from the methionine-specific tRNA used for peptide chain elongation. Therefore, all proteins synthesized in eukaryotic cytosol start with this amino acid, unless it is clipped in course of posttranslational modifications. However, polypeptides synthesized by mitochondria and chloroplasts start with N-formylmethionine.

---

*However, fluorophores at the amino end of a nascent peptide may affect its movement through the ribosome [67].

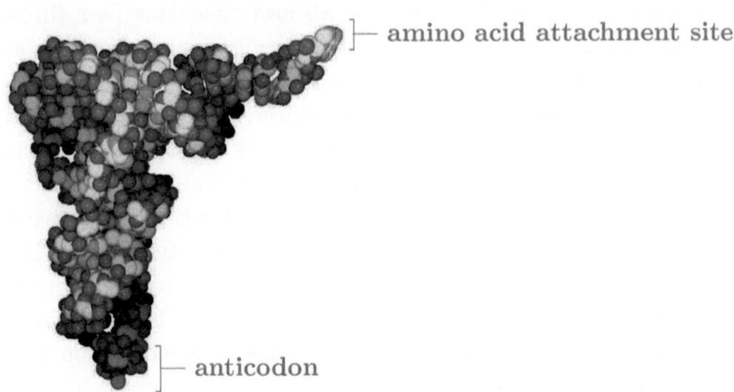

FIG. 1.39: The structure of the phenylalanine-specific tRNA from *Saccharomyces cerevisiae*, determined by E. Westhof *et al.* [68] by means of X-ray crystallography. Atomic coordinates are obtained from the RCSB Protein Data Bank record 4TRA. This record does not contain positions of hydrogen atoms, therefore they are not shown. The molecule is visualized using SIVIPROF.

At the initial stage, the **A** and **E** sites are blocked by initiation factors, which also prevent preliminary association of ribosomal subunits. Besides, the conformation of the mRNA at the **E** site in an initiation complex does not allow base paring. Therefore, only the **P** site with the start codon is available for binding a tRNA.

After subunit association, the initiation factors are released and the **A** site accommodates a charged tRNA with the corresponding anticodon. The binding is assisted by an elongation factor. The accuracy of the base-pairing is checked by the decoding element on the A-site of the small subunit. A correct codon-anticodon match causes a cascade of conformational rearrangements, which distribute from the small to the large subunit. As a result, the carboxyl group of the amino acid that was bound to the tRNA at the **P** site is transferred by the catalytic site of the large subunit to the amino group of the amino acid that is linked to the tRNA at the adjacent **A** site, with formation of a new peptide bond. This reaction is termed *transpeptidation*. After that, the ribosome moves to the next codon, to accept a new tRNA with a corresponding amino acid. The tRNA holding the carboxyl end of the growing peptide chain is now situated at the **P** site. The deacylated tRNA is ejected from the **E** site upon accommodation of a new tRNA on the **A** site (see Fig. 1.40) The amino end of the nascent chain presumingly passes through the tunnel in the large subunit (Fig. 1.38(e,f)) as the peptide grows.

The process of peptide elongation is repeated until the ribosome reaches a stop codon, signaling the termination of synthesis. The stop codons are recognized by release factors, which catalyze the hydrolysis of the ester bond between the tRNA and the nascent peptide. This event is followed by release of the mRNA, the tRNAs and dissociation of the ribosome into subunits.

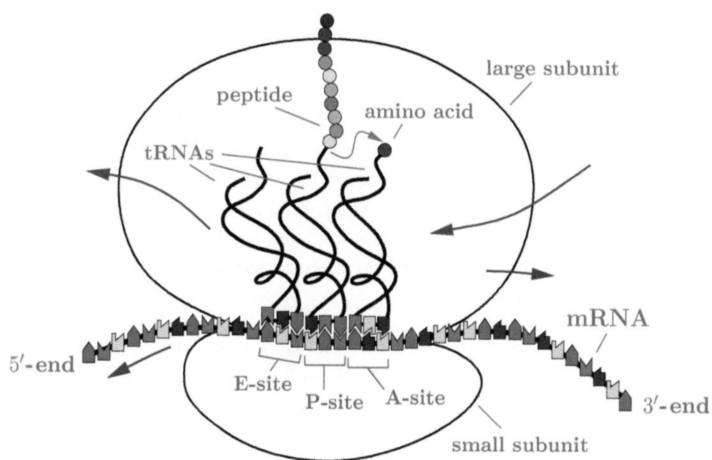

FIG. 1.40: A schematic representation of a translation process. Green, blue and dark red arrows show the directions of transpeptidation, tRNA exchange and relative movements of the mRNA or the ribosome respectively. To the moment, most of diagrams found in literature either depict only **A** and **P** sites, or show the **E** site without base pairing on mRNA. However, recent results of crystallographical and microbiological investigations approve the existence of the codon-anticodon interaction at the **E** site [69, 70] and its importance for prevention of reading frame shifting* [71].

1.8.3 Cotranslational protein folding

The specificity and the accuracy of rearrangements aimed for correct positioning of substrates in PTC assume that the amino acids are forced to acquire certain standard conformations before and after transpeptidation.

As already mentioned in Subsection 1.6.1, the dihedral angle ϕ of proline residues is fixed at values about $-60°$. The corresponding favorable values for ψ approximately lay in the ranges from $-100°$ to $0°$ and from $100°$ to $180°$ (see Fig. 1.33). The former range includes the region of an α-helical structure, while the latter is more typical for turns or for the structure that is adopted in collagen. Presumably, one of this configurations is adopted universely by all residues at PTC, providing the initial conformation for subsequent folding.

V. I. Lim and A. S. Spirin [43, 72] have analyzed the stereochemistry of ribosomal transpeptidation and came to the conclusion that the residue directly attached to the tRNA at the **P** site must have ϕ and ψ dihedral angles equal to $-60°$, and

---

*Since there is no separation signs between codons, unintentional shifting the reading frame by one or two nucleotides results in incorrect interpretation of all subsequent code.

that the transpeptidation reaction yields the same dihedral angles for the appended residue. Such dihedral angle values are close to those in an α-helical conformation[*], which, in principle, can be adopted by any amino acid. Therefore, it has been suggested [43, 72] that the nascent peptide is passed into the ribosomal tunnel in the form of an α-helix. This conformation should be suitable for pushing through the tunnel, since it is more rigid and less adhesive to hydrophilic tunnel walls (due to its self-saturation with hydrogen bonds) than an extended one. Moreover, it would be favored by space limitations in the tunnel and by fixation of the carboxyl end in the same configuration. After emerging from the ribosomal tunnel, the helix could be destabilized and subsequently rearranged into the unique three-dimensional structure of the native form.

Based on exploration of later elicited ribosomal crystal structures, A. Bashan *et al.* [73, 74] have proposed a model for peptide bond formation machinery. The authors concluded, in contrast to the findings of V. I. Lim and A. S. Spirin, that a rotatory motion of the amino acid attachment segment of the tRNA at the **A** site, resulting in peptide bond formation, would ensure that peptides are directed to the ribosomal tunnel in an extended β-like conformation [75].

C. DasGupta *et al.* [76–78] have shown that 50S ribosomal subunits of *Escherichia coli* are able to refold a large number of various denaturated proteins. In experiments with rhodanese by W. Kudlicki *et al.* [79] (partially reviewed in [49]), about 10 % of the denaturated enzyme could refold spontaneously under used conditions, while in presence of 50S ribosomal subunits this value increased to nearly 70 %, and reached 97 % when chaperones were additionally supplied. Renaturation achieved almost 100 % when rhodanese was incubated with the elongation factor EF-G and GTP in addition to ribosomes. Moreover, it has been observed that ribosome-assisted folding of cytoplasmic malate dehydrogenase is distinctly different from independent renaturation of this enzyme [80]. The folding function has been traced to the domain V of the 23S rRNA, which constitutes the PTC. Two segments of the domain V act consequently: one binds the protein and alters its conformation, and the other one aids its release. The large rRNA of plants and animals exhibits the same activity. This function could be inherited from the catalytic structural features. However, it has been shown that certain nucleotides of the PTC interact with specific amino acids[†] during refolding [78]. It was presumed that some proteins may start folding at the PTC, modulated by its activity, and then leave the ribosome via the interface between the two subunits.

What is the conformation enforced at the PTC exit, and whether it is universal or sequence dependent, remains to be clarified. It is also possible that the dynamic

---

[*]SiViProF simulations show that if this conformation remains unchanged upon appending further amino acids, the resulting periodic structure is an unstable right-handed helix, which readily rearranges into an α-helix.

[†]Most interactions were detected for asparagine, but some were also observed for lysine, glutamine, leucine, and glycine. No interactions were found for negatively charged amino acids.

tunnel makes selective contributions to the folding. In any case, it is very likely that proteins start folding into their native form not from a random arrangement, but rather from a certain configuration facilitating fast and correct folding.

This hypothesis is supported by the fact that folding of nascent proteins *in vivo* is noticeably faster than refolding *in vitro* from a denaturated quasi-random structure [46–49]. In fact, the latter process is often too slow to be utilized in a cell. The observed time of protein folding *in vivo* is of order 10^{-1}–10^3 seconds [5]. Renaturation of a simple monomeric enzyme ribonuclease under optimal conditions of dilution, pH and temperature takes about 20 minutes, while multidomain proteins may require several hours [48]. The half-lives of eukaryotic proteins vary from about 30 seconds to many days [54], depending on protein functions. Intracellular enzymes involved in metabolic pathways have short lifetimes, enabling finer regulation of metabolism according to cell needs. Proteins that fail to fold sufficiently fast are marked for rapid degradation.

Moreover, the rate of protein synthesis suggests that nascent chains can begin to acquire their native configuration cotranslationally, starting from the amino end. The prokaryotic translation rate in a substrate-rich medium at 37° C is about 15 to 20 amino acids per second. In a poorer medium or in a cell-free translation system, the rate of only about 10 codons per second is achieved. The average time of the elongation cycle is about 3 times longer at 25° C [43]. The translation rate in eukaryotes can achieve 10 amino acids per second at 37° C, but usually it is lower and can vary significantly due to presence of control mechanisms that can pause translation. A typical duration of an eukaryotic elongation cycle is in the range from 0.1 to 0.5 seconds [43].

Thus, the average time of the elongation cycle in both prokaryotes and eukaryotes normally exceeds 0.05 seconds, implying that at least 5 to 25 seconds are required to synthesize a typical globular monomeric protein consisting of 100–500 residues. The synthesis of such a protein in eukaryotes can take more than 10 times longer.

The dimensions of the tunnel do not leave much freedom to form a tertiary structure. However, formation of a helix or its unfolding into an extended form inside the tunnel could be possible. The amino end emerging from the tunnel can immediately start more complex conformational rearrangements, while the carboxyl end is fixed at the PTC.

Already since the sixties it has been known that about 30 to 40 C-proximal amino acid residues remain protected from proteinases in crude ribosome preparations [43, 81–83]. As mentioned in the above given references, this number of residues would span around 50 Å if the protected fragments were in the α-helical conformation, or more than 100 Å in a completely extended one. The length of the tunnel is approximately 100 Å. This fact could imply that the protected chains were in extended conformations, as concluded by many researchers. However, in the discussed experiments, described in [81, 82], the ribosomes with labeled peptides were centrifuged before or during the proteolysis, and this could result in stretching of possibly initially helical chains inside the ribosomal tunnels.

In addition to that, a number of experiments have shown that much shorter fragments can be accessible to digestion (see [43], p. 372, and references therein). These results, in addition to those described in Subsection 1.8.1, could indicate that not all polypeptides enter the ribosomal tunnel.

T. Tsalkova *et al.* [84] (also reviewed in [49]) examined the interaction of anti-coumarin antibodies with coumarin that was incorporated in the amino terminus of bovine rhodanese, bacterial chloramphenicol acetyltransferase (CAT), and MS2 coat protein. The results demonstrated that short peptides were unreactive. The accessible for antibodies longer chains were of different size and contained different number of residues, suggesting that peptides may acquire different conformations inside the tunnel. For example, accessible CAT peptides achieved about 8.5 kDa, or 72 amino acid residues, which should extend to 108 Å being in an α-helical conformation. MS2 coat protein fragments were recognized by antibodies when they achieved a size of about 4.5 kDa, containing 44 residues. Analysis of crystal structures of CAT and MS2 coat protein revealed that the former primarily contains helices, while the latter tends to form β-sheets [49]. Earlier, similar studies of W.D. Picking *et al.* (reviewed in [49, 84]) have shown that polyalanine peptides, which have high propensity to form α-helices, became accessible to antibodies and their antigen binding fragments when the nascent chains achieved 60 or 50 residues respectively.

C. A. Woolhead *et al.* [85] (also reviewed in [83]) used fluorescence resonance energy transfer (FRET) to measure the distance between two fluorescent dyes incorporated in a nascent chain with a separation of 24 residues. The obtained results suggest that the conformations of a transmembrane sequence (TMS) and fragment belonging to a soluble secretory protein are drastically different inside the tunnel. The latter fragment apparently moves through the tunnel in an extended conformation, while the TMS acquires a more compact structure with the coefficient of energy transfer similar to the one of an α-helix. This structure is observed immediately in the vicinity of the PTC and remains essentially constant until the TMS reaches a ER membrane (see Subsection 1.8.4 for information about targeting of transmembrane and secretory proteins). However, this helical structure unfolds upon entering the cytosol in the absence of translocation agents. Moreover, it was observed that some ribosomal proteins are likely to interact with the TMS, but not with the secretory protein. The authors concluded that the folding of the TMS is induced by the ribosome on a sequence-selective basis and that the resulting compact conformation of the TMS is not stable by itself in an aqueous solution, but rather stabilized in the membrane hydrophobic environment.

Yet, another possible explanation for the described effects could be that the elements of the secondary structure start forming according to energetic preferences of the given amino acid sequence immediately upon its entering from the PTC to the tunnel, possibly favored by the local ionic environment of the latter, with one or another initial conformation preset at PTC. The observed interactions with ribosomal proteins inside the tunnel may be a consequence of the formed structure, and not necessary its determinant. Destabilization of helical structures with exposed hydrophobic patches in cytoplasm could be due to interaction with chaperones (see Subsection 1.7.3).

Numerous examples show that proteins can acquire their functional tertiary structure cotranslationally, while their carboxyl end is still fixed at the PTC (see, for example, reviews in [43, 49]), and even associate with free protein subunits to form an active quaternary structure (reviewed in [43]). Although protein synthesis in prokaryotic cells usually proceeds at higher rates, there are evidences that cotranslational folding occurs in both eukaryotes and prokaryotes (reviewed in [49]).

Taking all together, many facts speak in favor of the consideration that ribosomes and the organization of protein synthesis contribute to a certain folding pathway, and these details may be important for successful modeling and simulation of protein folding.

1.8.4 Posttranslational modifications

Emerging polypeptide chains can be subjected to further chemical processing. It includes cleavage of polypeptide fragments, attachment of certain chemical groups, and formation of disulfide bonds between cysteine residues. In particular, many transmembrane and secretory proteins require glycosylation.

Although proteins are synthesized in cytosol (with exception of some mitochondrial proteins), most of the posttranslational modifications occur in the ER, possibly with subsequent transformations in the Golgi apparatus. Proteins, targeted to the ER, contain leading signal sequences, usually consisting of 25 ± 11 amino acids [86]. When a signal sequence emerges out of the ribosomal tunnel, a signal recognition particle suspended in cytosol arrests the elongation of the polypeptide chain until the ribosome binds to a membrane translocation channel, also called *translocon*, of the ER. Then the synthesis is resumed, and the polypeptide passes to the ER lumen or stays anchored in the membrane (in case of a transmembrane protein). A signal peptidase, located near the translocation pore, cleaves the signal sequences of secretory and some transmembrane proteins.

Additionally, some proteins require cleavage of other parts to become biologically active. For example, to produce mature insulin, a central fragment of proinsulin is removed by specific proteases in the Golgi apparatus, while the terminal parts of the initial polypeptide remain connected by two disulfide bridges.

In eukaryotic cells, disulfide bonds are mainly formed in the ER lumen, favored by its oxidative environment. They are typical for secretory, lysosomal, and some membrane proteins. Formation and breakage of disulfide bonds is catalyzed by protein disulfide isomerase, which thereby assists in protein folding. In the reducing environment of cytosol disulfide bonds are generally unstable.

In the ER lumen, where proteins are present in high concentrations, a polypeptide has to be protected by chaperones from the aggregation with other folding chains, until it acquires a proper three-dimensional structure (see also Subsection 1.7.3). Certain types of glycosylation are essential for correct folding and start cotranslationally as the polypeptide is passed into the ER lumen [48]. When the protein is modified and folded as necessary, it is transported to the Golgi apparatus for further processing, sorting, and delivery to final destinations. Misfolded proteins are identified in the ER, marked, and retranslocated to cytosol for degradation by proteasomes.

1.9   EXPERIMENTAL PROTEIN STRUCTURE DETERMINATION

Experimentally determined native structures are a valuable source of information that can be used for evaluation of computational structure prediction, as well as for analysis of native structural features and respective model improvement. It also serves as a foundation for knowledge-based structure prediction approaches. However, utilizing experimental results, it is important to take into account the quality of the data and to be aware of limitations of each particular method of experimental structure determination.

1.9.1   X-RAY CRYSTALLOGRAPHY*

The resolving power of a microscope is limited by the wavelength of the visible light, which is in the range from 380 to 750 nm. Objects smaller than half the wavelength can not be seen. X-rays have a wavelength from 0.7 to 1.5 Å, and can resolve a molecular structure, being scattered by electrons and captured by a photographic film or detector device. However, a reconstruction of the macromolecular configuration from the diffraction pattern is a complex problem. The number of atoms in a repeating unit of protein crystals achieves thousands or even hundreds of thousands, and each of them makes contribution to the whole diffraction pattern, not to a single spot on it. More precisely, the reflection image is related to the electron density via Fourier transform, but the information about phase angles is lost and only the reflection intensity is recorded. Besides, compared to crystals of small molecules, protein crystals contain unusually large amount of solvent, which often results in less order and gives a blurred image.

The process of an X-ray structural analysis includes the following steps. First, a protein must be purified and crystallized in a way that its structure still maximally resembles its native form. Obtaining a proper crystal of a sufficient size, which

---

*See, for example, [87] (or [54]) for more information.

should be typically at least $0.2 - 0.3$ mm in each dimension [87], can be a difficult task. Native structures of many proteins are not yet explored by this reason.

Then the crystal is placed in an X-ray beam, in order to obtain its diffraction patterns. Data have to be collected as quickly as possible, since biomolecules deteriorate in the intense radiation. Analysis of measurements includes correction of known systematic errors and determination of the accuracy by averaging reflections that should be identical due to symmetry, with a subsequent evaluation of the data consistency.

The phase angles, necessary for determination of the electron-density map, are not known. An initial estimate for phase angles is made, for example, by comparison with a known similar structure or by supplementing the crystal with electron-dense metal atoms and observing consequent changes in scattering. The estimated phase angles together with the available diffraction patterns are used for computation of an approximate electron-density map. If the resolution and the initial guess for phases are sufficiently good, one can recognize fragments of secondary structure and roughly classify the amino acid side chains by size. A molecular model is built to fit the obtained electron density. The corresponding diffraction pattern and possible attenuation of scattering, caused by thermal motion, are evaluated and used for model refinement. The process is repeated until the investigator is convinced that there is no evidence of a significant inconsistency and no further improvement is expected.

The vast majority of experimental data about native structures is obtained by X-ray crystallography. Although the environment of the protein in a crystal is different from physiological, a comparison with structures determined by other methods shows that conformation in a crystal generally represents the native form of the protein [54], apart from conformations of the side chains on the protein surface, influenced by packing in the crystal. Sometimes naturally unstable chains may appear more ordered due to interaction with surrounding groups. The electron density of hydrogen atoms is relatively week and often can not be restored. Therefore, records of determined structures usually do not contain hydrogen positions. Sometimes the positions that were used in model refinement are published.

1.9.2 Nuclear magnetic resonance (NMR) spectroscopy[*]

This technique for protein structure determination has become available in the middle eighties, about thirty years later than the X-ray crystallography [87]. A big advantage of the NMR spectroscopy is that it is not restricted to crystallizable macromolecules. It permits investigation of proteins in a natural environment and even observation of their dynamics [54].

Certain atoms have a nuclear spin that can be detected by NMR spectroscopy. Among those are ^{1}H, naturally abundant in proteins, as well as rare isotopes ^{13}C

---

[*]A more detailed description can be found in [87].

and ^{15}N, which can be embedded instead of common ^{12}C and ^{14}N with a purpose of investigation. When a strong static magnetic field is applied, magnetic dipoles produced by nuclear spins align themselves either parallel or antiparallel to the external field. The former orientation corresponds to a state of a low energy, and the latter is a high energy state. Then a short pulse of radio waves in a resonance frequency range is given in a perpendicular direction, which makes some nuclei to absorb energy and transit to a higher energy state. The separation of these energy states is very small compared to those of electronic, translational and rotational states [87]. The signals are subtle, therefore the results of numerous such experiments are averaged to increase signal-to-noise ratio. The thereby obtained absorption spectrum contains information about the nuclei and their immediate surrounding.

A disposition of electronegative atoms or delocalized electrons in the neighborhood results in a shift of the resonance frequency. Besides, nuclear spins of nearby atoms or protons bound to covalently bonded atoms interact in certain detectable ways. These effects provide information about interatomic distances and connectivity, although an extraction of this data may require an *a priori* knowledge of the molecular structure and give ambiguous results due to spectral overlaps. One of the main difficulties of spectra interpretation is the identification of atoms giving each particular signal.

The large amount of hydrogen atoms in proteins makes one-dimensional ^{1}H NMR spectra too complicated. Even the analysis of two-dimensional spectra, obtained by later developed techniques, is a complex process, including a construction of a molecular model and its subsequent refinement. Spectral analysis can be facilitated by heteronuclear NMR experiments, when rare isotopes ^{13}C and ^{15}N are feeded for protein synthesis. For some experiments proteins are dissolved in ^{2}H$_2$O, since otherwise the response of solvent can strongly dominate.

A number of optimized three-dimensional structures, consistent with distance constrains from NMR spectroscopy, is generated utilizing the knowledge about amino acid sequence, chirality, atomic van der Waals radii, bond lengths and angles. Strictly speaking, the real native structure does not have to be among the obtained models, since the conformational space is not sampled completely [87]. Usually the whole ensemble of the computed structures is published (see Figure 1.41). Due to averaging of multiple spectra, rigid chains give stronger signals, resulting in more constrains and respectively less ambiguity. Therefore, when the images of the obtained models are superimposed, these parts of the molecule tend to coincide, whereas some other fragments show more variance in conformations, giving insight into chain flexibility and its thermal motion. However, discrepancies can arise from experimental uncertainties of another kind, such as overlapping resonance frequencies. Since some interactions of nuclear spins decay with the growing distance, obtained molecular models may be biased in an unpredictable way towards structures with less separation of such nuclei [87].

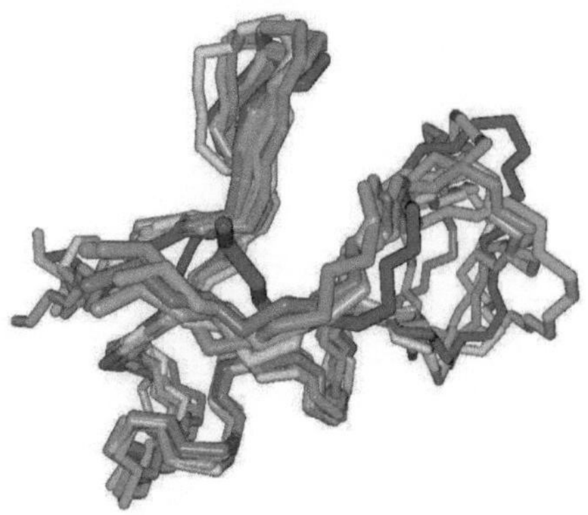

FIG. 1.41: The main chain conformations for seven models of the chlorotoxin solution structure determined by G. Lippens *et al.* [88] by means of NMR spectroscopy (RCSB Protein Data Bank record 1CHL) and visualized by SiViProF.

1.10 STRUCTURE DATABASES

One of the most important resources providing information about known protein structures is the *RCSB Protein Data Bank* (PDB), which can be accessed via Internet under the address *http://www.rcsb.org* or *http://www.pdb.org*. PDB is a part of the Worldwide Protein Data Bank (wwPDB), a union of organizations for deposition, processing, and distribution of structural data for biological macromolecules, their fragments and complexes.

PDB contains more than fifty thousand records, and their number grows with an increasing speed. PDB records contain experimentally determined atomic coordinates, as well as information about the method of structure determination and its accuracy. It also provides some tools for visualization of molecular structures and data analysis. Besides, the structural information can be downloaded and processed using the available PDB file format description.

Chapter 2

Approaches for Protein Structure Prediction

2.1 From *ab initio* to knowledge-based approaches

There is an immense number of literature sources related to various methods for molecular modeling and protein structure prediction. The volume of probably any book written on this subject exceeds the number of pages that can be devoted to the literature review in this thesis. Therefore, we shall only have a quick glance at the approximate research directions, popular in this area, and then focus on the issues particularly related to this work.

The methods for protein structure prediction can be roughly subdivided into groups related to *ab initio* and knowledge-based approaches, which can be also combined with each other. However, due to the complexity of the problem, up to date neither one of them, nor their combination, has proven to give in general reliable predictions based only on amino acid sequence. *Ab initio* methods are usually very demanding computationally or contain prohibitive simplifications. Thus, the potential of such approaches grows with the raising computational power of computers. The deficiencies of knowledge-based predictions, on the other hand, are typically ascribed to the shortage of known structures. Therefore, further substantial improvement is expected with an accelerating increase in the number of experimentally determined folds. The prediction techniques are evaluated each two years in a community-wide competition, called CASP*, where preliminary unknown target structures are determined by X-ray crystallography or NMR spectroscopy, after submission of predictions by currently more than 150 research groups [89].

2.1.1 *Ab initio* protein folding

Ab initio methods make their predictions based on physical laws. Quantum mechanical computations are not feasible for large molecules or molecular ensembles. For treatment of such systems, empirical *force fields* were introduced, which describe

---

*CASP stands for "Critical Assessment of techniques for protein Structure Prediction".

either the potential energy of the system or the potential of mean force acting at the considered protein molecule (both further referred as *the potential U*) as a function of positions of interacting atoms. Depending on the system features and application area, these models can differ in complexity. However, many of them have common terms, discussed later in this chapter.

Among *ab initio* approaches are *molecular dynamics* methods, which aim to describe successive configurations of the system using equations of motion. Besides, there are various ways of minimizing the potential, which look only for stable states.

Molecular dynamics methods are diverse, and their usage is not restricted to the protein folding problem. They are based on statistical thermodynamics and operate by concepts like canonical ensembles. A *canonical ensemble*, also referred as NVT, can be understood as a closed system of a specified composition (N particles), volume V, and temperature T, replicated a large number of times [14]. In a *microcanonical ensemble* (NVE), the number of particles N, the volume V, and the energy E are specified. Other ensembles can be designed in a similar way.

A common problem of the classical molecular dynamics is that it requires time steps of the order about 10^{-15} s. This results in fast accumulation of integration errors with respect to the time scales of interest. Therefore, special methods must be designed to minimize this drawback. Besides, for a closed system, simple integration of motion equations represents sampling from a microcanonical ensemble, where the energy is preserved. This produces artifacts like artificial heating of solvent molecules, which accompanies a decrease of the protein energy. To reproduce the experimental conditions, it is usually desired to sample from NTP, and not from NVE. Therefore, the development of methods designed for sampling from ensembles different from the microcanonical have obtained much attention [90, 91].

As already discussed in Subsections 1.7.1 and 1.7.2, it is important to take the interactions with the solvent into account. However, all-atom continuous molecular dynamics with explicit inclusion of water molecules requires an immense computational effort. The simulation box must be chosen large enough to eliminate the artifacts related to the boundary conditions*. This implies a significant increase in the number of atoms participating in simulations. Even on a supercomputer, a simulation of only early stages of protein folding takes months. Probably the longest simulations of this kind were done in the project *Folding@Home*, involving world wide distributed computing. These simulations still stay at maximum in a millisecond range [92], whereas a real folding process may take several minutes.

This problem is sometimes partially circumvented by using implicit solvent models. In this case, the equations of motion are supplemented by a friction term and a stochastic term[†] aiming to reproduce random forces that arise due to collisions with

---

*Normally it is desired to implement periodic boundary conditions for better representation of the bulk solvent.

[†]A more detailed discussion of this approach can be found in the review by A. Neumaier [5].

solvent atoms. However, this description is rather simplified and not devoid of drawbacks. For example, the matrix of friction coefficients is modeled as a scalar multiple of the mass matrix, despite the fact that not all atoms are in contact with the solvent.

Another approach based on statistical thermodynamics is the *Metropolis Monte Carlo* method. It can be used separately or combined with molecular dynamics to obtain hybrid methods [91]. The idea of the Metropolis method is to generate a sequence of random configurations, such that the transition to each new state satisfies the following conditions. A new state is accepted if it corresponds to a lower energy, but rejected with a certain probability if the energy raises. Thus, a random number between 0 and 1 is generated and compared with the Boltzmann factor f_B for the difference ΔG between the energies of the new and current configuration:

$$f_B = e^{-\Delta G/k_B T}. \tag{2.1}$$

If the random number is larger than the computed Boltzmann factor, the transition is rejected [20]. As explained in the first chapter, for molar energies the gas constant $\bar{R}$ should be used instead of the Boltzmann's constant k_B. The deficiency of the classical Metropolis Monte Carlo method is that there is no control for existence of a low energy transition path between each two subsequent states. Hence, a transition to a lower energy state would be always accepted, even if the configurations are virtually separated by a high energy barrier. Combination with other methods may help to eliminate this drawback.

Both Monte Carlo and molecular dynamics simulations may stick for a long time in a low energy valley, if the probability to escape is sufficiently low. To speed up the sampling of the diverse regions in the conformational or phase space, the *simulated annealing** is used. This technique consists in heating up the system, followed by slow cooling. Often this is done with the purpose to find the global energy minimum, which is assumed to correspond to the native state. Some weakness of this strategy can be seen in the fact that in nature heating normally causes denaturation. Thus, starting with a near-native state one can arrive to a completely incidental structure.

Apart from the simulated annealing, various other approaches aiming to find the global minimum of energy have been proposed†. Among them are smoothing techniques, which blur out numerous local minima in the original potential function by an artificial diffusion. Alternatively, they start with a simplified model and introduce subsequently more details for the energy computations. Besides, some genetic methods are used, which imitate evolutional crossing over and mutations for generation of new low-energy candidates. Branch and bound methods are used for an exhaustive search over discretized conformational spaces. A similar strategy is sometimes applied for simple lattice models, in which the chain of amino

---

*See, for example, the reviews in [5, 20] for more information.

†A more detailed review of these methods is given in [5].

acid residues is represented by a sequence of hydrophobic and hydrophilic beads, restricted to adopt only discrete lattice positions.

The basis for all global minimization methods is the assumption that the native fold of a protein represents the conformation corresponding to the global minimum of energy. This can be seen as a possible critique point for these approaches. As already discussed in Subsections 1.7.3-1.7.4 and pointed out by A. Neumaier in the earlier review [5], the hypothesis of the global minimum can be disputed. Thus, for a long polypeptide chain, a wide variety of conformations is possible. Among them may exist metastable states, separated by so large energy barriers that a transition between those conformations is rather improbable within the protein turn-over time at physiological temperature. Therefore, for biological tasks it may be sufficient to preset during the protein translation an initial conformation favoring fast correct folding of the protein into its native form.

Besides, a search for the global minimum on a rough energy surface with a large amount of local minima is a computationally demanding process with a questionable quality of results, owing to the inaccuracy of energy estimations.

2.1.2 KNOWLEDGE-BASED STRUCTURE PREDICTION

Knowledge-based approaches use the information about already known structures to predict a likely native arrangement of a new sequence. In this category are statistical and machine learning methods for secondary structure prediction, which focus mainly on the dihedral angles of the main chain and usually do not take into account the interaction of remote sequence fragments. Besides, there are tertiary structure prediction techniques, such as homology modeling and threading, which use known folds as templates to reconstruct the structure of a new sequence.

Among the first knowledge-based methods were those proposed by P.Y. Chou and G.D. Fasman [36], as well as by J. Garnier, D. J. Osguthorpe, and B. Robson [93]. The Chou-Fasman method makes secondary structure predictions based on the α-helical and β-sheet propensities of the individual amino acid residues (see Table 2.1). These propensities were estimated from statistical analysis of 15 proteins with known structures and yielded 77 % accuracy in prediction, whether the residue belongs to a helix, β-sheet, or coil, for the other tested 19 proteins [36]. The idea of the method is to search among six-residue segments for the α-helix and β-sheet nucleation regions, such that certain conditions, involving the conformational parameters P_α and P_β, as well as the α-helical or β-sheet assignments from Table 2.1, are fulfilled. After that, the α-helices and β-sheets are extended in both directions until one of the specified termination criteria is satisfied.

Some other authors brought out the opinion that the rules stated in the Chou-Fasman method are "open to interpretation" and "have not always yielded such promising results in the hands of other workers" [93]. In any case, the values in Table 2.1 certainly convey useful information: the observed tendencies are not

TABLE 2.1: The assignment of the amino acid secondary structure propensities based on the P_α and P_β conformational parameters, adopted from [36]*.

Helical assignments			β-sheet assignments		
Residues	P_α	**Assignment**	**Residues**	P_β	**Assignment**
glutamic acid	1.53	H_α	methionine	1.67	H_β
alanine	1.45	H_α	valine	1.65	H_β
leucine	1.34	H_α	isoleucine	1.60	H_β
histidine	1.24	h_α	cysteine	1.30	h_β
methionine	1.20	h_α	tyrosine	1.29	h_β
glutamine	1.17	h_α	phenylalanine	1.28	h_β
tryptophan	1.14	h_α	glutamine	1.23	h_β
valine	1.14	h_α	leucine	1.22	h_β
phenylalanine	1.12	h_α	threonine	1.20	h_β
lysine	1.07	I_α	tryptophan	1.19	h_β
isoleucine	1.00	I_α	alanine	0.97	I_β
aspartic acid	0.98	i_α	arginine	0.90	i_β
threonine	0.82	i_α	glycine	0.81	i_β
serine	0.79	i_α	aspartic acid	0.80	i_β
arginine	0.79	i_α	lysine	0.74	b_β
cysteine	0.77	i_α	serine	0.72	b_β
asparagine	0.73	b_α	histidine	0.71	b_β
tyrosine	0.61	b_α	asparagine	0.65	b_β
proline	0.59	B_α	proline	0.62	b_β
glycine	0.53	B_α	glutamic acid	0.26	B_β

completely accidental but result from the amino acid properties. At least some of these propensities can be readily explained, as discussed in [1].

In the basic, or directional, version of the Garnier-Osguthorpe-Robson (GOR) method, the amino acid structural propensities are treated as conditional, depending on the next 16 neighbors (i.e. eight preceeding and eight following the considered residue) in the chain. The authors of the GOR method have specified four matrices of the size 20×17, one for each of the following conformational types: an α-helix, an extended conformation like a β-sheet, turns, and a coil. Each row in the matrix

---

*H_α – strong α-helix former, h_α – α-helix former, I_α – weak α-helix former, i_α – α-helix-indifferent, b_α – α-helix breaker, B_α – strong α-helix breaker, H_β – strong β-sheet former, h_β – β-sheet former, I_β – weak β-sheet former, i_β – β-sheet-indifferent, b_β – β-sheet breaker, B_β – strong β-sheet breaker. Despite the assignments given in the table, proline and aspartic acid residues are placed to the I_α-category when they are near the helical N-terminal, while tryptophan residues are treated as b_β, when they are near the C-terminal of a β-sheet.

corresponds to a certain type of amino acid residues, and each column stands for a position in the chain from $j-8$ to $j+8$, where j is the number of the residue, for which the conformation is to be defined.

The value $m_{ik}^{[s]}$ ($s = \overline{1,4}$, $i = \overline{1,20}$, $k = \overline{1,17}$) in a matrix $\mathbf{M}^{[s]}$ represents a measure of the likelihood* of the central, i.e. j-th, residue to be found in a specific conformation s under the condition that the residue of the type corresponding to the i-th row is on the position $j-9+k$, which is related to the k-th column. The central column contains the unconditional propensities of residues for the considered conformation. The likelihood $I^{[s]}$ of each conformational type s for the j-th residue is estimated as the sum of 17 directional information measures selected from the matrix $\mathbf{M}^{[s]}$ according to the amino acid sequence of the surrounding segment of 17 residues, starting at the $(j-8)$-th residue. Finally, the conformation s with the highest likelihood value $I^{[s]}$ is chosen. To improve the prediction results, a decision constant specific for each conformation type can be subtracted from $I^{[s]}$ before comparison.

In the single-residue information version of the GOR method, only the information for each residue about its own propensity for the given conformation is considered. However, also the average propensity of the local chain segment including this residue is taken into account. The segment is not necessarily centered about the considered residue, but has to be shifted in a way that the maximal average propensity is obtained. The length of the segment is given by the *run constant* n_s, which can have a different value for each particular type of the secondary structure. Optimal values for n_s were found to be 6 for a helix, 5 for an extended conformation, 4 for a reverse turn, and 3 for a coil, in agreement with the rules for the nucleation and termination of α-helices and β-sheets in the Chou-Fasman method.

Both versions of the GOR method could predict correctly about 48-49 % of residue conformations (considered on the described four-state basis) in a test with 26 proteins [93].

Homology modeling is based on the fact that proteins having a common ancestor, and therefore certain sequence similarity, often adopt similar folds. In the simplest cases of the protein design, when a point mutation is introduced with a purpose of potential stabilization of an enzyme, one can use the atomic coordinates from the known structure of the original enzyme, replace the side chain of the amino acid that was changed, and compare the energies of the two structures, to see whether the replacement was favorable. Undoubtedly, the utilization of this scenario is not limited by a single substitution. Moreover, it has been shown that the fold may remain the same even when the proteins share only 30 % of the sequence identity [40]. Typically, in homologous proteins the fragments having a regular secondary structure are better conserved, while replacements, insertions, and deletions mostly

---

*Both, "measure" and "likelihood" are used here as abstract concepts, and not in the sense of homonymous mathematical notions.

occur in the loop and coil regions [40], where they have less impact on the overall structure. Replacements are usually more easy to treat than the other modifications, in particular, if the amino acid is substituted by another one having similar properties. The side chain similarity is reflected in a related sequence match score. However, due to insertions and deletions, comparison of sequences involves a complex fragment alignment process, which is still not always performed in an optimal way.

Threading methods are used when no sufficiently close homologue with a known structure can be found. They appeared due to observations that even proteins having no statistically significant sequence similarity may have the same fold type and that the number of possible folds appears limited, governed by certain rules. Consequently, the idea of threading is to verify whether the given amino acid sequence is likely to adopt any of the fold types from a library of fold templates. The scoring strategies to evaluate a match can be very different. Some of them are based on amino acid properties or involve prediction of the secondary structure.

To the moment, the most successful approaches for protein structure prediction use a combination of very different methods, starting from a search for a homologous protein with a known structure, then, particularly if no good template homologue is found, make a model of the secondary structure and use it for selection of compatible fold templates in threading. Finally, energy minimization and molecular dynamics methods are utilized to get rid of chain clashes, to model the sequence insertions, and to distinguish good candidates from obviously non-native folds. Force fields including solvation energy terms have been shown to be particularly useful for the latter selection process [40].

2.2 EMPIRICAL FORCE FIELDS

In order to formalize the mathematical description of the problem and to set a basis for its further mathematical treatment, let us introduce a new system of notations.

Let a unique number be assigned to each atom, and let $\mathcal{N}$ be the set of atom numbers for a given molecule. Let $N := \mathrm{card}\,\mathcal{N}$ denote the number of atoms. In consistence with notations introduced in Section 1.6.1, let us refer to the atom with the number i as A_i, denote by $\vec{r}_i \in \mathbb{R}^3$ the vector pointing from the coordinate origin to the center of atom A_i, and set $\vec{r}_{ij} := \vec{r}_j - \vec{r}_i$, $i,j \in \mathcal{N}$. Further, let us denote by $\angle(\vec{a}, \vec{b})$ the angle between any vectors $\vec{a}$ and $\vec{b}$, and refer to the dihedral angle $\angle(A_i - A_j - A_k - A_l)$ as $\angle(\vec{r}_{ij}, \vec{r}_{jk}, \vec{r}_{kl})$, in order to emphasize its dependence on instant atomic coordinates.

Let the expression $i \bowtie j$ signify that the atoms A_i and A_j are joined by a covalent bond, and $i \nsim j$ denote that A_i and A_j are bonded neither directly nor to a common neighbor. The both relations imply that $i \neq j$. Besides, let us utilize the expression $i \simeq j$ to indicate that atoms A_i and A_j are connected by a double or a partially double bond. Note that $i \simeq j$ implies $i \bowtie j$, but not vice versa.

Most of the popular models for molecular simulations are based on an estimation of the potential energy arising from the electrostatic and van der Waals interactions between atoms and, apart from that, contain terms introducing energy punishments for deviations of bond lengths, bond angles, and some torsion angles from their equilibrium states (see, e.g., [94–98] or the reviews in [20, 99, 100]). Additionally, they may include energy punishments for the out-of-plane bending of atoms bound to aromatic rings.

Using the introduced notations, we can write a classical version of a model from the described above category as follows:

$$
U(\vec{\mathbf{r}}_{12}, \vec{\mathbf{r}}_{13}, \ldots, \vec{\mathbf{r}}_{N-1\,N}) = \sum_{\substack{i,j \in \mathcal{N} \\ i \bowtie j, \, i<j}} k_{ij}^{[b]} (\|\vec{\mathbf{r}}_{ij}\| - l_{ij})^2 + \sum_{\substack{i,j,k \in \mathcal{N} \\ i \bowtie j \bowtie k, \, i<k}} k_{ijk}^{[a]} (\angle(\vec{\mathbf{r}}_{ji}, \vec{\mathbf{r}}_{jk}) - \alpha_{ijk})^2 +
$$

$$
+ \sum_{\substack{i,j,k,l \in \mathcal{N} \\ i \bowtie j \bowtie k \bowtie l \\ i<l}} \sum_{m=0}^{M} k_{ijklm}^{[d]} (1 + \cos(m\angle(\vec{\mathbf{r}}_{ij}, \vec{\mathbf{r}}_{jk}, \vec{\mathbf{r}}_{kl}) - \zeta_{ijklm})) + \sum_{\substack{i,j,k,l \in \mathcal{N} \\ i \bowtie j, \, i \simeq k, \, i \simeq l, \, k<l}} \frac{k_i^{[o]} (\vec{\mathbf{r}}_{ij} \cdot (\vec{\mathbf{r}}_{ik} \times \vec{\mathbf{r}}_{il}))^2}{\|\vec{\mathbf{r}}_{ik} \times \vec{\mathbf{r}}_{il}\|^2} +
$$

$$
+ \sum_{\substack{i,j \in \mathcal{N} \\ i \nsim j, \, i<j}} k_{ij1}^{[w]} \left(\left(\frac{k_{ij2}^{[w]}}{\|\vec{\mathbf{r}}_{ij}\|} \right)^{12} - \left(\frac{k_{ij2}^{[w]}}{\|\vec{\mathbf{r}}_{ij}\|} \right)^{6} \right) + \frac{1}{4\pi\epsilon_0} \sum_{\substack{i,j \in \mathcal{N} \\ i \nsim j, \, i<j}} \frac{q_i q_j}{\|\vec{\mathbf{r}}_{ij}\|}, \quad (2.2)
$$

where $k_{ij}^{[b]}$, l_{ij}, $k_{ijk}^{[a]}$, α_{ijk}, $k_{ijklm}^{[d]}$, ζ_{ijklm}, $k_i^{[o]}$, $k_{ij1}^{[w]}$, $k_{ij2}^{[w]}$ are parameters depending on the type and hybridization of the atoms with the corresponding numbers, ϵ_0 is the vacuum permittivity, and q_i denotes the partial charge of atom A_i. The meaning of separate terms in equation (2.2) is discussed in the following subsections.

The last term can be scaled to take into account the polarization of the medium. However, not all force field developers do that. Consideration of the screening effect is particularly important in cases of implicit solvent representation.

Some models include a term to describe hydrogen bonding, which can be modeled in various ways. A few of them are discussed in Subsection 2.2.4. Besides, other terms can be added to account for the hydrophobic effect and electric free energy of solvation, as discussed in Section 2.3.

2.2.1   HARMONIC APPROXIMATION FOR BOND LENGTHS AND ANGLES

The first two terms in equation (2.2) introduce energy punishments for deviations of bond lengths and bond angles from their equilibrium states. l_{ij} and α_{ijk} denote respectively the most favorable bond length and angle for the corresponding atoms. Minimization of a potential function containing only these two terms is sufficient for reconstruction of the shape of some simple molecules, such as methane or benzene (Fig. 2.1), starting from random atomic positions.

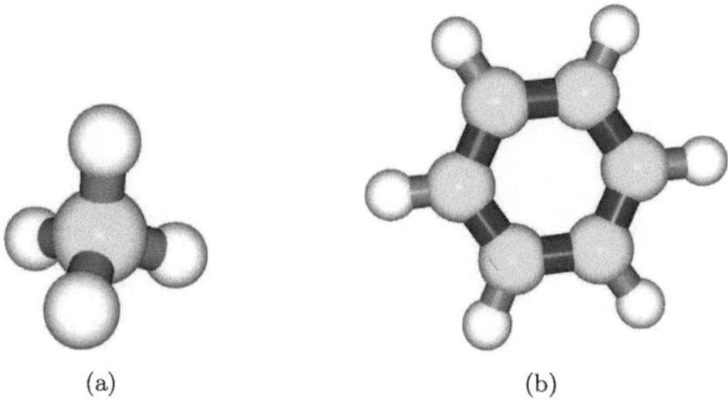

FIG. 2.1: Molecules of methane (a) and benzene (b).

The given harmonic approximation can be improved by introducing higher order and cross terms or even replacing it by a Morse potential (see, e.g., the reviews in [5, 20]), in order to obtain better prediction of vibrational frequencies. However, as already mentioned in Section 1.6, the conformation of a polypeptide is mainly described by torsion along the chain, whereas the constrains for bond lengths and angles are needed rather for the purpose of fixing their values close to the optimal ones. Therefore, in the following we shall not consider such kind of extensions.

2.2.2 Torsion angles and out-of-plane bending

Quantum mechanical computations suggest the existence of energy barriers for rotation of chemical bonds. For example, the energy of the ethane molecule is maximized in the eclipsed (Fig. 2.2 (a)) and minimized in the staggered (Fig. 2.2 (b)) conformation. To capture this effect, many force fields contain terms of the form

$$U_{ijkl}^{[d]} = \sum_{m=0}^{M} k_{ijklm}^{[d]}(1 + \cos(m\angle(\vec{r}_{ij}, \vec{r}_{jk}, \vec{r}_{kl}) - \zeta_{ijklm})) \qquad (2.3)$$

or

$$U_{ijkl}^{[d]} = \sum_{m=0}^{M} k_{ijklm}^{[d]} \cos(\angle(\vec{r}_{ij}, \vec{r}_{jk}, \vec{r}_{kl}))^{m}. \qquad (2.4)$$

The *multiplicity* m in equation (2.3) gives the number of minima, and ζ_{ijklm} is the *phase factor*, which determines their location. For example, the potential due to rotation of a double bond, where the related dihedral angle is restricted to the values about 0 and π, can be represented by a single term with $m = 2$ and $\zeta_{ijklm} = \pi$.

In some cases inclusion of terms describing the van der Waals interactions may be sufficient to model the repulsion of atoms at the opposite ends of a dihedral angle. Nevertheless, torsional terms are necessary for the modeling of non-rotatable double and partially double bonds, if the values of the corresponding dihedral angles are not constrained in some other way.

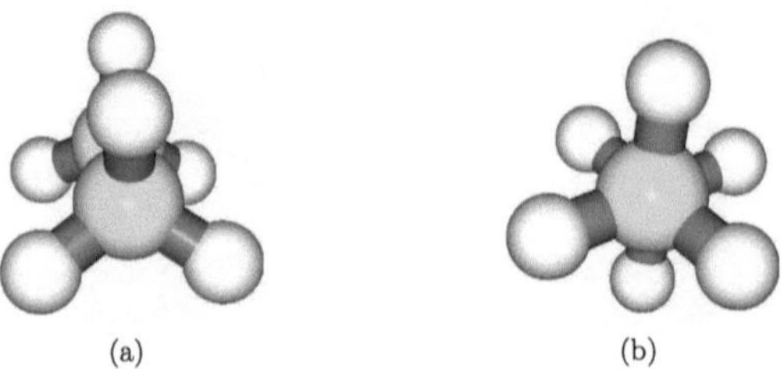

(a)   (b)

FIG. 2.2: Molecules of ethane in the eclipsed (a) and staggered (b) conformation.

A punishment term $U^{[o]}$ for the *out-of-plane bending* may be necessary for computation of some ring structures. For example, consider the imidazole ring depicted in Figure 2.3. The reference values for the bond angles in sp^2-hybridized nitrogen and carbon atoms would cause the hydrogens to deviate from the plane, whereas it is known that π-bonding results in a coplanar arrangement.

Sometimes the out-of-plane bending is represented in terms of *improper torsion angles*, i.e. angles $\angle(A_i - A_j - A_k - A_l)$ where the atoms A_i, A_j, A_k and A_l are not connected consequently, but instead, atom A_i, for example, is bound to the other three. The potential term to enforce an improper torsion angle ζ to take values about 0 or π is then given by

$$U^{[o]}(\zeta) = k^{[o]}(1 - \cos(2\zeta)).$$

However, usually $U^{[o]}$ is specified in one of the following forms:

$$U^{[o]}(h) = k^{[o]}h^2 \tag{2.5}$$

or

$$U^{[o]}(\theta) = k^{[o]}\theta^2, \tag{2.6}$$

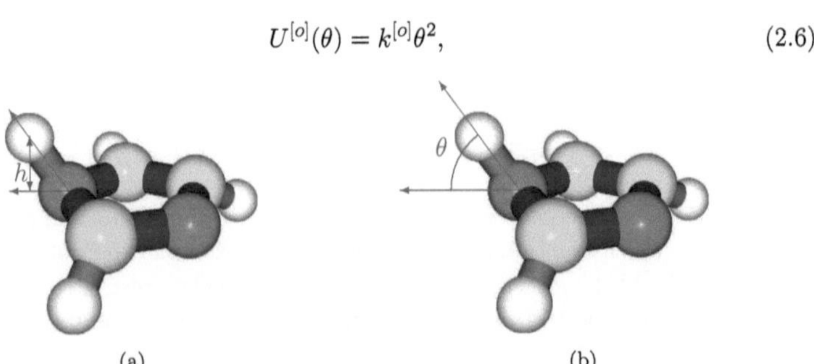

(a)   (b)

FIG. 2.3: The out-of-plane bending in the imidazole ring. (a) h is the distance between the center of the considered out-of-plane atom and the plane of the ring. (b) θ is the angle between the direction of the related bond and the ring plane. Note that according to quantum mechanics all atoms in this structure should lay in one plane.

where h or θ are defined as shown in Figure 2.3, $k^{[o]}$ is a constant depending on atom types. Using the introduced notations, we can write the contribution of all out-of-plane bending terms of the form (2.5) as

$$\sum_{\substack{i,j,k,l \in \mathcal{N} \\ i \bowtie j,\, i \simeq k,\, i \simeq l,\, k < l}} \frac{k_i^{[o]} (\vec{\mathbf{r}}_{ij} \cdot (\vec{\mathbf{r}}_{ik} \times \vec{\mathbf{r}}_{il}))^2}{\|\vec{\mathbf{r}}_{ik} \times \vec{\mathbf{r}}_{il}\|^2},$$

and of the form (2.6) as:

$$\sum_{\substack{i,j,k,l \in \mathcal{N} \\ i \bowtie j,\, i \simeq k,\, i \simeq l,\, k < l}} k_i^{[o]} \arcsin^2 \left(\frac{\vec{\mathbf{r}}_{ij} \cdot (\vec{\mathbf{r}}_{ik} \times \vec{\mathbf{r}}_{il})}{\|\vec{\mathbf{r}}_{ij}\| \|\vec{\mathbf{r}}_{ik} \times \vec{\mathbf{r}}_{il}\|} \right).$$

Clearly, the second form is computationally more demanding.

2.2.3 NON-BONDED INTERACTIONS AND ASSIGNMENT OF ATOM CHARGES

The last two terms in equation (2.2) stand for the van der Waals and electrostatic interactions. Alternative treatments of non-bonded interactions also exist. For example, in the intermolecular force field proposed by D. E. Williams [101], the energy $U_{ij}^{[n]}$ arising from these interactions between any two atoms A_i and A_j from different molecules is described as follows[*]:

$$U_{ij}^{[n]} = a_{ij} e^{-b_{ij} \|\vec{\mathbf{r}}_{ij}\|} - c_{ij} \|\vec{\mathbf{r}}_{ij}\|^{-6} + \frac{q_i q_j}{4\pi\epsilon_0 \|\vec{\mathbf{r}}_{ij}\|},$$

where a_{ij}, b_{ij}, and c_{ij} are given by

$$a_{ij} = \sqrt{a_i a_j}, \qquad b_{ij} = 0.5(b_i + b_j), \quad \text{and} \quad c_{ij} = \sqrt{c_i c_j}$$

for specified parameters a_i, b_i, and c_i ($i \in \mathcal{N}$) depending on atom types and classes of functional groups, to which these atoms belong.

In the general purpose Tripos 5.2 force field [94], the Lennard-Jones potential of the form (cf. (1.42))

$$U_{ij}^{[w]} = \sqrt{E_i^{[w]} E_j^{[w]}} \left(\left(\frac{R_i^{[w]} + R_j^{[w]}}{\|\vec{\mathbf{r}}_{ij}\|} \right)^{12} - 2 \left(\frac{R_i^{[w]} + R_j^{[w]}}{\|\vec{\mathbf{r}}_{ij}\|} \right)^6 \right) \tag{2.7}$$

is used for the modeling of the van der Waals interactions. Here $E_i^{[w]}$ and $R_i^{[w]}$ ($\forall i \in \mathcal{N}$) are respectively a specified energy constant and the atomic van der Waals radius, both depending on the atom type. The same term is adopted to describe van der Waals interactions in this work, together with the associated and some other parameters of the Tripos 5.2 force field.

---

[*]In the original version, the factor $1/4\pi\epsilon_0$ is absent, apparently indicating the usage of different energy units.

The term for the electrostatic energy in the Tripos 5.2 force field involves a constant or distance dependent dielectric function. Besides, for distances smaller than 0.5 Å a linear extrapolation is used for prevention of a numerical overflow. The evaluation of electrostatic interactions, however, was omitted by the authors of the Tripos 5.2 force field during its testing. The reasons for that were ambiguities in the definition of partial charges and dielectric functions.

Indeed, the discrepancy in values of charges computed by different methods can be of an order of magnitude for some atoms, and even the sign of charges can change [20, 27, 102]. Point partial charges are an approximation, which is aimed to reproduce electrostatic properties of atom nuclei with surrounding electron clouds. Clearly, any kind of a point approximation for the electron density would inevitably fail at sufficiently short distances. However, apart from this limitation there is a problem related to determination of the best point charge model and criteria for its evaluation.

The effect of an electron density shift towards certain nuclei is detectable experimentally, for example, through the resulting electric dipole moments (see Subsection 1.5.3). The latter can be used as a criterion for the evaluation of the quality of a partial charge distribution and may also permit unambiguous assignment of partial charges in the simplest cases. For instance, a molecule of HF with the bond length of 0.917 Å has the dipole moment of 1.82 D, which can be obtained with the opposite charges of $\pm 0.413 \, |e^-|$ placed at the two nuclei [20]. However, already for somewhat larger molecules there is often more than one possible charge distribution that can yield the given dipole moment. Besides, for molecules with a non-fixed geometry, the measured dipole moment represents the average for the forms that are present in the experimental sample. This poses an additional problem not only for determination of partial charges, but also for their assessment based on experimental values of the dipole moments.

One way to assign partial charges is to determine from quantum mechanical computations the electron density and to partition it between atom nuclei by one of the known methods, such as Mulliken or Löwdin population analysis (see [20] and references therein). Another approach is to compute from the obtained electron density the electrostatic potential in different points around the molecule and choose the charge values that reproduce it in the best way. However, the computed electron density depends on the chosen basis set for molecular orbitals and conformation of the molecule. Besides, the results for different point sets may differ significantly. Moreover, placing charges aside atom centers can give a considerably better fit (see, for example, the review in [20]).

A relatively fast procedure for assignment of partial charges, which is adopted in this work, was proposed by J. Gasteiger and M. Marsili [26, 27]. The idea of the method is to shift a certain amount of negative charge from atoms with lower electronegativity values to more electronegative atoms. The authors of the procedure

adopted the definition of the orbital electronegativity, which was introduced by
J. Hinze [23–25] as given by equation (1.41). However, their interpretation of this
concept is somewhat different, since they do not distinguish between σ-bonding
and π-bonding orbitals during charge computations (see Algorithm 2.1). Rather,
the electronegativity is treated as a characteristic of an atom in a certain valence
state.

Despite that disagreement, J. Gasteiger and M. Marsili speculate that the electronegativity of each atomic orbital must depend on the total charge of the atom
and approximate this dependence by a three-term truncation of a Maclaurin series:

$$\chi_i^{[k]}(q_i) = a_i^{[k]} + b_i^{[k]} q_i + c_i^{[k]} q_i^2. \tag{2.8}$$

Here i and k are the atomic and the orbital number respectively, $a_i^{[k]}$, $b_i^{[k]}$, and $c_i^{[k]}$
are the corresponding parameters, and q_i is the total charge of the atom.

Using the electronegativity definition given by equation (1.41), J. Gasteiger and
M. Marsili have obtained the coefficients $a_i^{[k]}$ from the published by J. Hinze and
H. H. Jaffé [23] values of the ionization potential $^0I_i^{[k]}$ and electron affinity $^0E_i^{[k]}$,
related to the orbitals of atoms in the uncharged state:

$$a_i^{[k]} = \chi_i^{[k]}(0) = \frac{1}{2}(^0I_i^{[k]} + {}^0E_i^{[k]}).$$

Likewise, they utilized the published in [25] values of the ionization potential $^+I_i^{[k]}$
and electron affinity $^+E_i^{[k]}$, related to atoms in cationic states with a unit charge:

$$a_i^{[k]} + b_i^{[k]} + c_i^{[k]} = \chi_i^{[k]}(1) = \frac{1}{2}(^+I_i^{[k]} + {}^+E_i^{[k]}).$$

Finally, they set the ionization potentials, related to negative ions, equal to the
electron affinities of atoms in the uncharged states and assumed that the electron
affinity of orbitals of a negative ion is zero. This yielded

$$a_i^{[k]} - b_i^{[k]} + c_i^{[k]} = \chi_i^{[k]}(-1) = \frac{1}{2}{}^0E_i^{[k]}$$

and finally defined the coefficients. However, no separate coefficients for different
orbitals were reported in [27], but one for each of the typical valence states of the
considered atoms.

The Partial Equalization of Orbital Electronegativity (PEOE) procedure can be
summarized in the form of Algorithm 2.1. The procedure is started with formal
charges, which are typically all zero for uncharged species. The authors of the
method report fast convergence and therefore recommend to stop the procedure
after 5-6 iterations. Step 2.1 must be performed for all atoms before proceeding to
the next step.

Algorithm 2.1 (J. Gasteiger and M. Marsili [26, 27])

1) Using the provided parameters compute cationic electronegativities for all atoms:

$$\chi_i(1) = a_i + b_i + c_i, \quad \forall i \in \mathcal{N},$$

2) repeat the following steps for 5-6 iterations:

2.1) determine the electronegativity for each atom at the n-th iteration:

$$(\chi_i)_n = a_i + b_i(q_i)_{n-1} + c_i \left(q_i^2\right)_{n-1}, \quad \forall i \in \mathcal{N},$$

2.2) compute the charge that each atom should obtain after n-th iteration:

$$(\Delta q_i)_n = \frac{1}{2^n} \left(\sum_{\substack{j \in \mathcal{N},\, j \bowtie i \\ (\chi_i)_n < (\chi_j)_n}} \frac{(\chi_j)_n - (\chi_i)_n}{\chi_i(1)} + \sum_{\substack{k \in \mathcal{N},\, k \bowtie i \\ (\chi_i)_n > (\chi_k)_n}} \frac{(\chi_k)_n - (\chi_i)_n}{\chi_k(1)} \right), \quad \forall i \in \mathcal{N}, \tag{2.9}$$

2.3) update the total charge for each atom:

$$(q_i)_n = (q_i)_{n-1} + (\Delta q_i)_n, \quad \forall i \in \mathcal{N}.$$

According to equation (2.9), the amount of charge transferred between each two bonded atoms is proportional to the difference in their electronegativities, which depend on the atom valence states and the charges obtained after the previous iteration. The electronegativity difference is scaled by the cation electronegativity of the less electronegative atom, in order to obtain a dimensionless quantity, which is interpreted as a portion of the unit charge.

The authors of the method insist, however, that full equalization of the electronegativity is unreasonable, since all atoms of the same type and valence state (for instance, all hydrogens in acetic acid) would obtain the same partial charge. Further, the authors speculate that the electrostatic field generated upon charge transfer must have an additional damping effect, which should hinder further electronegativity equalization. Therefore, they introduce a heuristic factor $1/2^n$ that reduces charge transfer and decreases with each iteration.

The argumentation used for justification of the method can be disputed, since the gradual electronegativity equalization by itself has an effect of damping further charge transfer. Besides, the proposed choice of the damping strategy has no sound physical basis.

However, the authors of the procedure found an excellent correlation between the computed charges and the C-1s binding energies obtained from ESCA* experiments, while a poor correlation was obtained with the charges from Mulliken population

---

*ESCA stands for the Electron Spectroscopy for Chemical Analysis

analysis [27]. Besides, J. Gasteiger and M. Marsili report a very good correlation of the PEOE hydrogen charges with ^{1}H NMR chemical shifts. Further, the authors observed a good linear correlation between the PEOE charges and the pK_a values of different compounds, ranging from an acidic HF to weak alkanes.

Simulations by SIVIPROF, performed in course of this work, show that the PEOE charges do not always reproduce experimentally determined dipole moments. A good fit is obtained for formaldehyde: the computed dipole moment is 2.30 D, while the experimental values collected from different sources and published in [103] vary from 2.17±0.02 D to 2.339±0.013 D. One of the references mentioned there gives the value 2.29 D. The computed values for formic acid change from about 1.29 D to 2.50 D, depending on the molecule conformation, while the published experimental values fall between 1.35 ± 0.02 D and 2.09 D. Computations for methanol give the dipole moment about 1.54 D, which is not very far from the minimal published values. The latter are in the range between 1.61 D and 3.10 D. For acetic acid the values about 1.61-1.67 D are predicted, while measurements give the dipole moments between 0.74 ± 0.2 D to 2.17 D. The computed dipole moment of methylamine is about 0.71 D, which is somewhat too small compared to the experimental values between 1.00 D and 1.47 D. Moreover, tests show that more complete electronegativity equalization yields apparently better dipole moments for σ-bonded systems (for instance, 2.12 D for methanol and 1.25 D for methylamine). However, the experimental values fall too much apart for a good reference.

It is interesting to note that if the PEOE charges would be, for instance, about 1.5 times smaller than the "correct" ones, possibly due to excessive damping of the electronegativity equalization, they would still yield a linear correlation with the C-1s binding energies, ^{1}H NMR chemical shifts, and pK_a values.

Despite the described objections, the PEOE procedure enjoys popularity due to its rapidness and hence good applicability to large molecules, as well as because other methods are also not free from drawbacks.

2.2.4 MODELS FOR HYDROGEN BONDING

In the most force fields it is assumed that the electrostatic and van der Waals interaction terms are sufficient to reproduce hydrogen bonding, therefore no explicit hydrogen bonding terms are included.

However, as discussed in Subsection 1.5.5, hydrogen bonds have a feature that the electronegative atom of the donor group is likely to approach the acceptor as close as if there would be no hydrogen atom between them. In other words, there is no typical van der Waals interaction between the hydrogen atom and the acceptor of the hydrogen bond at short distances. For this reason, in the general purpose Tripos 5.2 force field [94] hydrogen atoms attached to potential donors of hydrogen bonds obtain zero van der Waals radius for their interactions with potential acceptors of hydrogen bonds.

Some other force fields include a term with distance-dependent energy contribution [104, 105] of the form:

$$U_{ha}^{[H]} = \frac{c_{ha}}{\|\vec{\mathbf{r}}_{ha}\|^{12}} - \frac{d_{ha}}{\|\vec{\mathbf{r}}_{ha}\|^{10}},$$

where c_{ha} and d_{ha} are specific coefficients, and $\|\vec{\mathbf{r}}_{ha}\|$ is the distance between the hydrogen atom and the acceptor of the hydrogen bond.

Sometimes, terms that take into account directionality of hydrogen bonding are incorporated [20]. For example, the MM3 force field includes a term of the form:

$$G_{ha}^{[H]} = \frac{k_{ha}^{[H]}}{\epsilon}\left(184000\, e^{-12\|\vec{\mathbf{r}}_{ha}\|/r_{ha}} - 2.25\cos\angle(\vec{\mathbf{r}}_{dh},\vec{\mathbf{r}}_{da})\frac{\|\vec{\mathbf{r}}_{hd}\|}{\ell_{hd}}\left(\frac{r_{ha}}{\|\vec{\mathbf{r}}_{ha}\|}\right)^{6}\right),$$

which describes directional hydrogen bonding [106, 107]. Here h, a, and d are the numbers related to the hydrogen atom, the acceptor, and the electronegative atom in the donor group, $k_{ha}^{[H]}$ is a specific parameter, ϵ is the dielectric constant, and r_{ha} is the optimal distance for the considered hydrogen bond.

2.3 Solvation models

We have discussed in the first chapter that an important factor, essentially determining the native structure of a protein, is the interaction with the surrounding solvent. Hydrophobic effect, in particular, is regarded by biochemists as a key player on early stages of protein globule formation. At the same time, charged groups, which interact with water dipoles, tend to stay on the surface.

Explicit inclusion of water molecules is a very demanding task from the computational point of view. A number of researches proposed implicit solvation models, based on estimation of the solvent exposed area or excluded volume [42, 108–114]. Usually it is assumed that the solvation energy for each atom is proportional to its instantaneous SASA, which is defined as described in Subsection 1.7.1:

$$\Delta G_i^{[s]} = k_i^{[s]} a_i^{[s]}(\vec{\mathbf{r}}_{12},\vec{\mathbf{r}}_{13},\ldots,\vec{\mathbf{r}}_{N-1\,N}), \quad i \in \mathcal{N}.$$

Here $k_i^{[s]}$ are atomic solvation parameters, and $a_i^{[s]}(\vec{\mathbf{r}}_{12},\vec{\mathbf{r}}_{13},\ldots,\vec{\mathbf{r}}_{N-1\,N})$ are the atomic areas exposed to the solvent. The evaluation of the latter, however, can be also computationally demanding.

2.3.1 Estimation of solvent-exposed area

Lee and Richards [42] estimated the static accessibility of atoms by sectioning the molecule structure by a set of parallel planes and summation of the approximate areas of segments confined between planes. Surface areas of segments they calculated based on the arc lengths of atom sections after elimination of the intersecting parts.

Shrake and Rupley [108] used a set of 92 fixed test points that were nearly uniformly distributed over each solvated atom sphere, in order to determine the atom exposure. The points were checked for an occlusion by test atoms through the comparison of the ratios of the solvated sphere radii of the test atoms to the distances from the centers of those atoms to the check point.

Connolly [109] utilized the definition of the molecular surface, introduced by Richards [115] as a part of the van der Waals surface accessible to a probe sphere, and presented a computer algorithm for analytical computation of the surface area by subdividing the surface into a set of pieces of spheres and tori.

Richmond [110] proposed an analytical approach for exact calculation of the SASA, by providing an expression for the surface area exterior to an arbitrary number of overlapping spheres.

Futamura *et al.* [111] presented a Monte Carlo algorithm for computing the SASA and corresponding error bounds. The authors also suggested sequential algorithms with parallelizations, in order to reduce the computational time for spherical intersection checking. These algorithms can be also used with other methods for SASA computation.

All those methods were implemented in different software packages and successfully used for SASA calculations, but their application for protein simulations is computationally very demanding due to a large number of atoms to be processed. More efficient numerical algorithms are normally less accurate.

Wodak and Janin [112] proposed a probabilistic analytical expression for fast approximation of the SASA and its partial derivatives relative to the distances between atoms. This approach is based on the assumption of randomly distributed atoms that are not allowed to penetrate each other. However, this condition does not hold for covalently bound atoms. Hasel *et al.* [113] and further Cavallo *et al.* [114] introduced some modifications of this method, which imply separate treatment of covalently bound and not bound atoms. These procedures involve some adjustable parameters optimized on a set of specifically chosen molecules. Although the predictions of SASAs given by the probabilistic approach were reported to be in a reasonably good agreement with the SASAs calculated using geometrical algorithms, it seems to be questionable whether one can expect reliable results in folding simulations of complex proteins.

2.3.2 POISSON-BOLTZMANN EQUATION

Another significant effect to be considered in the protein modeling is the dielectric screening. Water has much higher electrical permittivity than the one of the hydrophobic core of proteins [1, 5, 17, 40] (see also Subsection 1.7.2). Hence, for realistic estimation of the electrostatic interactions in cases of an implicit solvent representation it is essential to take into account whether the charges are located on the surface of the protein globule or in the interior. However, in the most force fields either a constant electrical permittivity is used, or a distance dependent dielectric function is introduced, which does not describe this effect adequately.

A more accurate way to evaluate the electrostatic component of the solvation free energy and to account for the screening effect is to solve the Poisson-Boltzmann equation in $\Omega \in \mathbb{R}^3$ [20]*:

$$\boldsymbol{\nabla} \cdot (\epsilon(\vec{\mathbf{r}})\boldsymbol{\nabla}\varphi_e(\vec{\mathbf{r}})) - \frac{N_A q_e^2 I}{500\epsilon_0 k_B T} \sinh(\varphi_e(\vec{\mathbf{r}})) = -\frac{\rho(\vec{\mathbf{r}})}{\epsilon_0}, \tag{2.10}$$

where $\varphi_e : \Omega \to \mathbb{R}$ is the Coulomb potential, I is the ionic strength of the solvent (measured in mol/l) and $\rho : \Omega \to \mathbb{R}$ is the charge density. Other notations are as described before.

The molar electric free energy of a charge q_i positioned in $\vec{\mathbf{r}}_i$ is then given by

$$G_i^{[e]} = N_A q_i\, \varphi_e(\vec{\mathbf{r}}_i).$$

Unfortunately, solving equation (2.10) on every energy evaluation step requires several hundred times more computational time [116]. Besides, the energy arising from the hydrophobic effect is not taken into account.

---

*In the original version the equation is written implying different units, which result in the factor 4π instead of $1/\epsilon_0$.

Chapter 3

Modeling Intracellular Protein Folding

3.1 The general idea of the new approach

As discussed already in Section 1.6, the process of protein folding consists mainly in twisting about single bonds, which is driven by electrostatic and van der Waals interactions (arising both between protein atoms and due to presence of solvent molecules around), as well as by the hydrophobic effect. However, the most terms in equation (2.2) specify geometry constrains, which are not strict but rather stiff for physiologically meaningful energy levels. Therefore, the function of these terms is mainly to reconstruct and sustain a reasonable geometry for small molecules, while for the protein modeling the corresponding energy contributions introduce additional complications by shading the driving effects. Besides, common molecular mechanics models do not take chirality into account.

The idea of the new approach is, on one hand, to reduce the classical molecular mechanics model by eliminating the geometry-constraining terms with the corresponding degrees of freedom and, on the other hand, to complement it by terms that can help to reproduce better the effects possibly determinant for protein folding pathways in living cells. The factors that are particularly important in this aspect were discussed in Chapter 1.

In the following, we shall describe the potential of mean force as a function of only the minimal number of non-fixed dihedral angles in a polypeptide chain. The molecule is thereby naturally constrained to preserve other geometry features. We can then generate reasonable amino acid geometries, such that all bond lengths, bond angles, and dihedral angles obtain their optimal values, and so that the other requirements, including chirality, are fulfilled. Besides, each amino acid shall be considered in an appropriate ionization state.

The amino acids shall be appended consequently in a way that the formed peptide group is disposed in the *trans* conformation, and each chain elongation step is to be followed by minimization of energy for the emerging protein fragment. To prevent the folding of the nascent chain around the carboxyl terminal before the synthesis is completed, the potential shall include a plane restriction term, which enforces folding in a given half-space.

To account for the interaction with water, we shall introduce an implicit solvation model that gives an estimate for the hydrophobic effect and electric free energy of hydration. Additionally, we shall take into account the polarization of the medium and the resulting dielectric screening, which is different for charges in the protein core and for those in contact with water. Furthermore, in cases of hydrogen bonding, the Lennard-Jones potential for the van der Waals interaction between the electronegative acceptor and the hydrogen of the donor group should be modified to allow a closer contact. Finally, we shall add terms aimed to describe formation of disulfide bridges.

Later on, we shall discuss twisting forces and the peculiarities arising from dynamics in the space of dihedral angles.

3.2   INTERATOMIC DISTANCES VERSUS DIHEDRAL ANGLES

By reasons described above, let bond angles and lengths be fixed at their optimal values, i.e.

$$\|\vec{\mathbf{r}}_{ij}\| = l_{ij}, \quad \forall i,j \in \mathcal{N}: \ i \bowtie j, \tag{3.1}$$

and

$$\angle(\vec{\mathbf{r}}_{ij}, \vec{\mathbf{r}}_{ik}) = \alpha_i \in (0, \pi), \quad \forall i,j,k \in \mathcal{N}: \ (i \bowtie j) \wedge (i \bowtie k). \tag{3.2}$$

We would like to explore, how the distance between atoms depends on torsion angles along the chain.

DEFINITION 3.1
Let the atoms bound to each other by a common bond be referred as *0–order-connected*, atoms bonded to the same atom be called *1-order-connected* and so on, with the *connection order* $\aleph_{ij}$ equal to the minimal number of consequently bonded atoms standing between two given atoms A_i and A_j in the chain.

DEFINITION 3.2
Let the dihedral angles that are customary used and sufficient for unambiguous specification of a polypeptide conformation be called the *primary dihedral angles* (for example, dihedral angles between 2-order-connected nitrogen or carbon atoms in a main chain, such as ϕ, ψ and ω dihedral angles, see Subsection 1.6.1), and let the corresponding atom sequence be referred as a *primary chain*. The *secondary dihedral angles*, which appear in cases of branching, such as the angles between nitrogen and oxygen atoms of the same residue in a main chain, can be derived from the primary dihedrals and bond angles.

In the following, we shall interpret the sign "$\doteq$" as "must be equal".

Proposition 3.3

Let A_i, A_j, A_k, A_l be the centers of four consequently bonded atoms as shown in Figure 1.20, and let

$$\zeta_{jk}^{il} := \angle(\vec{\mathbf{r}}_{ij}, \vec{\mathbf{r}}_{jk}, \vec{\mathbf{r}}_{kl}) \quad \text{and} \quad \vec{\mathbf{e}}_{mn} := \frac{\vec{\mathbf{r}}_{mn}}{\|\vec{\mathbf{r}}_{mn}\|}, \ \forall\, m, n \in \mathcal{N}$$

here and further in the text. Then, for given $\vec{\mathbf{e}}_{ij}, \vec{\mathbf{e}}_{jk}$, and fixed bond angles, the direction $\vec{\mathbf{e}}_{kl}$ can be expressed in a related orthonormal basis, such as

$$\vec{\mathbf{e}}_{jk}, \quad \vec{\mathbf{e}}_{ijk} := \frac{\vec{\mathbf{e}}_{ij} \times \vec{\mathbf{e}}_{jk}}{\|\vec{\mathbf{e}}_{ij} \times \vec{\mathbf{e}}_{jk}\|}, \quad \vec{\mathbf{e}}_{(i)jk}^{\perp} := \vec{\mathbf{e}}_{jk} \times \vec{\mathbf{e}}_{ijk}, \tag{3.3}$$

or

$$\vec{\mathbf{e}}_{ij}, \quad \vec{\mathbf{e}}_{ijk}, \quad \vec{\mathbf{e}}_{ij(k)}^{\perp} := \vec{\mathbf{e}}_{ij} \times \vec{\mathbf{e}}_{ijk}, \tag{3.4}$$

as a function of the dihedral angle ζ_{jk}^{il}, namely:

$$\vec{\mathbf{e}}_{kl} = \vec{\mathbf{e}}_{ijk} \sin \alpha_k \sin \zeta_{jk}^{il} - \vec{\mathbf{e}}_{(i)jk}^{\perp} \sin \alpha_k \cos \zeta_{jk}^{il} - \vec{\mathbf{e}}_{jk} \cos \alpha_k, \tag{3.5}$$

or

$$\vec{\mathbf{e}}_{kl} = \vec{\mathbf{e}}_{ijk} \sin \alpha_k \sin \zeta_{jk}^{il} + \vec{\mathbf{e}}_{ij(k)}^{\perp}(\sin \alpha_j \cos \alpha_k + \cos \alpha_j \sin \alpha_k \cos \zeta_{jk}^{il}) +$$
$$+ \vec{\mathbf{e}}_{ij}(\cos \alpha_j \cos \alpha_k - \sin \alpha_j \sin \alpha_k \cos \zeta_{jk}^{il}) \tag{3.6}$$

respectively.

Proof. First of all, note that

$$\angle(\vec{\mathbf{e}}_{ji}, \vec{\mathbf{e}}_{jk}) = \alpha_j,$$

therefore

$$\angle(\vec{\mathbf{e}}_{ij}, \vec{\mathbf{e}}_{jk}) = \pi - \alpha_j \quad \text{and} \quad \vec{\mathbf{e}}_{ij} \cdot \vec{\mathbf{e}}_{jk} = \cos(\pi - \alpha_j) = -\cos \alpha_j. \tag{3.7}$$

Besides, we have:

$$\vec{\mathbf{e}}_{ijk} = \frac{\vec{\mathbf{e}}_{ij} \times \vec{\mathbf{e}}_{jk}}{\sin(\pi - \alpha_j)} = \frac{\vec{\mathbf{e}}_{ij} \times \vec{\mathbf{e}}_{jk}}{\sin \alpha_j}, \quad \text{and } \vec{\mathbf{e}}_{(i)jk}^{\perp} = \frac{\vec{\mathbf{e}}_{jk} \times (\vec{\mathbf{e}}_{ij} \times \vec{\mathbf{e}}_{jk})}{\sin \alpha_j}.$$

Since

$$\vec{\mathbf{e}}_{jk} \times (\vec{\mathbf{e}}_{ij} \times \vec{\mathbf{e}}_{jk}) = \vec{\mathbf{e}}_{ij}(\vec{\mathbf{e}}_{jk} \cdot \vec{\mathbf{e}}_{jk}) - \vec{\mathbf{e}}_{jk}(\vec{\mathbf{e}}_{jk} \cdot \vec{\mathbf{e}}_{ij}) = \vec{\mathbf{e}}_{ij} + \vec{\mathbf{e}}_{jk} \cos \alpha_j,$$

it follows that

$$\vec{\mathbf{e}}_{(i)jk}^{\perp} = \frac{\vec{\mathbf{e}}_{ij} + \vec{\mathbf{e}}_{jk} \cos \alpha_j}{\sin \alpha_j}. \tag{3.8}$$

Reciprocally

$$\vec{\mathbf{e}}_{ij(k)}^{\perp} = \frac{\vec{\mathbf{e}}_{ij} \times (\vec{\mathbf{e}}_{ij} \times \vec{\mathbf{e}}_{jk})}{\sin \alpha_j} = \frac{\vec{\mathbf{e}}_{ij}(\vec{\mathbf{e}}_{ij} \cdot \vec{\mathbf{e}}_{jk}) - \vec{\mathbf{e}}_{jk}(\vec{\mathbf{e}}_{ij} \cdot \vec{\mathbf{e}}_{ij})}{\sin \alpha_j} = \frac{-\vec{\mathbf{e}}_{ij} \cos \alpha_j - \vec{\mathbf{e}}_{jk}}{\sin \alpha_j}. \tag{3.9}$$

Let

$$\vec{e}_{kl} = a\,\vec{e}_{ijk} + b\,\vec{e}^{\,\perp}_{(i)jk} + c\,\vec{e}_{jk} \tag{3.10}$$

be a possible representation of $\vec{e}_{kl}$ in the basis (3.3). This decomposition is expected to satisfy the equalities

$$\angle(\vec{e}_{jk}, \vec{e}_{kl}) \doteq \pi - \alpha_k, \tag{3.11}$$

and

$$\angle(\vec{e}_{ij}, \vec{e}_{jk}, \vec{e}_{kl}) \doteq \zeta^{il}_{jk}. \tag{3.12}$$

Therefore, it should hold:

$$\vec{e}_{jk} \cdot \vec{e}_{kl} \doteq -\cos\alpha_k \tag{3.13}$$

and

$$\mathrm{sign}((\vec{e}_{jk} \times \vec{e}_{kl}) \cdot \vec{e}_{ij})\arccos \frac{\vec{e}_{ijk} \cdot (\vec{e}_{jk} \times \vec{e}_{kl})}{\sin\alpha_k} \doteq \zeta^{il}_{jk}, \tag{3.14}$$

see (1.44)-(1.46).

Substitution of (3.10) into (3.13) yields:

$$\vec{e}_{jk} \cdot (a\,\vec{e}_{ijk} + b\,\vec{e}^{\,\perp}_{(i)jk} + c\,\vec{e}_{jk}) = c \doteq -\cos\alpha_k. \tag{3.15}$$

Utilizing (3.8) and (3.10) we deduce:

$$\vec{e}_{jk} \times \vec{e}_{kl} = a\,\vec{e}_{jk} \times \vec{e}_{ijk} + b\,\vec{e}_{jk} \times \vec{e}^{\,\perp}_{(i)jk} = a\,\vec{e}^{\,\perp}_{(i)jk} + b\,\vec{e}_{jk} \times \frac{\vec{e}_{ij} + \vec{e}_{jk}\cos\alpha_j}{\sin\alpha_j} =$$

$$= a\,\vec{e}^{\,\perp}_{(i)jk} + b\,\frac{\vec{e}_{jk} \times \vec{e}_{ij}}{\sin\alpha_j} = a\,\vec{e}^{\,\perp}_{(i)jk} - b\,\vec{e}_{ijk}. \tag{3.16}$$

Substitution of (3.16) into (3.14) results in:

$$\mathrm{sign}(a\,\vec{e}^{\,\perp}_{(i)jk} \cdot \vec{e}_{ij})\arccos\left(\frac{-b}{\sin\alpha_k}\right) \doteq \zeta^{il}_{jk},$$

therefore

$$b \doteq -\sin\alpha_k \cos\zeta^{il}_{jk} \tag{3.17}$$

and

$$\mathrm{sign}(a\,\vec{e}^{\,\perp}_{(i)jk} \cdot \vec{e}_{ij}) \doteq \mathrm{sign}\,\zeta^{il}_{jk}. \tag{3.18}$$

The latter equation yields after substitution of (3.8):

$$\mathrm{sign}\left(\frac{a(\vec{e}_{ij} + \vec{e}_{jk}\cos\alpha_j) \cdot \vec{e}_{ij}}{\sin\alpha_j}\right) = \mathrm{sign}\left(\frac{a(1 - \cos^2\alpha_j)}{\sin\alpha_j}\right) =$$

$$= \mathrm{sign}(a\sin\alpha_j) \doteq \mathrm{sign}\,\zeta^{il}_{jk},$$

meaning that

$$\mathrm{sign}(a) \doteq \mathrm{sign}\,\zeta^{il}_{jk}, \tag{3.19}$$

since $\alpha_j \in (0, \pi)$.

Besides, $\vec{e}_{kl}$ must be a unit vector, therefore using (3.15) and (3.17) we obtain:

$$|a| = \sqrt{1 - b^2 - c^2} = \sqrt{1 - \sin^2 \alpha_k \cos^2 \zeta_{jk}^{il} - \cos^2 \alpha_k} =$$
$$= \sqrt{\sin^2 \alpha_k \sin^2 \zeta_{jk}^{il}} = |\sin \alpha_k \sin \zeta_{jk}^{il}|. \quad (3.20)$$

Since $\zeta_{jk}^{il} \in (-\pi, \pi]$ and $\alpha_k \in (0, \pi)$, equations (3.19) and (3.20) imply that

$$a = \sin \alpha_k \sin \zeta_{jk}^{il}. \quad (3.21)$$

It is obvious from the given derivation that the representation (3.5) in the basis (3.3) is unique, and therefore there exists only one unit vector with the properties (3.11) and (3.12).

Now we show that (3.6) provides the representation of $\vec{e}_{kl}$ given by (3.5) in the basis (3.4). Since (3.3) and (3.4) are orthonormal bases, the following relation holds:

$$\vec{e}_{kl} = \vec{e}_{ijk}(\vec{e}_{ijk} \cdot \vec{e}_{kl}) + \vec{e}_{ij(k)}^{\perp}(\vec{e}_{ij(k)}^{\perp} \cdot \vec{e}_{kl}) + \vec{e}_{ij}(\vec{e}_{ij} \cdot \vec{e}_{kl}), \quad (3.22)$$

and from (3.5) immediately follows:

$$\vec{e}_{ijk} \cdot \vec{e}_{kl} = \sin \alpha_k \sin \zeta_{jk}^{il}. \quad (3.23)$$

Using (3.8) and (3.9) we obtain:

$$\vec{e}_{ij} \cdot \vec{e}_{(i)jk}^{\perp} = \frac{\vec{e}_{ij} \cdot (\vec{e}_{ij} + \vec{e}_{jk} \cos \alpha_j)}{\sin \alpha_j} = \frac{1 - \cos^2 \alpha_j}{\sin \alpha_j} = \sin \alpha_j,$$

$$\vec{e}_{ij(k)}^{\perp} \cdot \vec{e}_{jk} = \frac{(-\vec{e}_{ij} \cos \alpha_j - \vec{e}_{jk}) \cdot \vec{e}_{jk}}{\sin \alpha_j} = \frac{\cos^2 \alpha_j - 1}{\sin \alpha_j} = -\sin \alpha_j, \quad (3.24)$$

$$\vec{e}_{ij(k)}^{\perp} \cdot \vec{e}_{(i)jk}^{\perp} = \frac{\vec{e}_{ij(k)}^{\perp} \cdot (\vec{e}_{ij} + \vec{e}_{jk} \cos \alpha_j)}{\sin \alpha_j} = \vec{e}_{ij(k)}^{\perp} \cdot \vec{e}_{jk} \frac{\cos \alpha_j}{\sin \alpha_j} = -\cos \alpha_j.$$

Consequently,

$$\vec{e}_{ij(k)}^{\perp} \cdot \vec{e}_{kl} = \vec{e}_{ij(k)}^{\perp} \cdot \vec{e}_{(i)jk}^{\perp}(-\sin \alpha_k \cos \zeta_{jk}^{il}) - \vec{e}_{ij(k)}^{\perp} \cdot \vec{e}_{jk} \cos \alpha_k =$$
$$= \cos \alpha_j \sin \alpha_k \cos \zeta_{jk}^{il} + \sin \alpha_j \cos \alpha_k \quad (3.25)$$

and

$$\vec{e}_{ij} \cdot \vec{e}_{kl} = \vec{e}_{ij} \cdot \vec{e}_{(i)jk}^{\perp}(-\sin \alpha_k \cos \zeta_{jk}^{il}) - \vec{e}_{ij} \cdot \vec{e}_{jk} \cos \alpha_k =$$
$$= -\sin \alpha_j \sin \alpha_k \cos \zeta_{jk}^{il} + \cos \alpha_j \cos \alpha_k. \quad (3.26)$$

Substitution of (3.23), (3.25), and (3.26) into (3.22) gives (3.6). $\square$

REMARK 3.4

Note that according to (3.9), the representation of $\vec{e}_{jk}$ in the basis (3.4) is:

$$\vec{e}_{jk} = -\cos \alpha_j \vec{e}_{ij} - \sin \alpha_j \vec{e}_{ij(k)}^{\perp}. \quad (3.27)$$

PROPOSITION 3.5

Under the conditions of Proposition 3.3 the distance between the centers of atoms A_i and A_l is given by

$$\|\vec{\mathbf{r}}_{il}\| = \sqrt{l_{ij}^2 + l_{jk}^2 + l_{kl}^2 - 2l_{ij}l_{jk}\cos\alpha_j - 2l_{jk}l_{kl}\cos\alpha_k +}$$
$$\overline{+2l_{ij}l_{kl}(\cos\alpha_j\cos\alpha_k - \sin\alpha_j\sin\alpha_k\cos\zeta_{jk}^{il})}. \quad (3.28)$$

Proof.

$$\|\vec{\mathbf{r}}_{il}\| = \sqrt{(\vec{\mathbf{r}}_{ij} + \vec{\mathbf{r}}_{jk} + \vec{\mathbf{r}}_{kl}) \cdot (\vec{\mathbf{r}}_{ij} + \vec{\mathbf{r}}_{jk} + \vec{\mathbf{r}}_{kl})} =$$
$$= \sqrt{\vec{\mathbf{r}}_{ij}\cdot\vec{\mathbf{r}}_{ij} + \vec{\mathbf{r}}_{jk}\cdot\vec{\mathbf{r}}_{jk} + \vec{\mathbf{r}}_{kl}\cdot\vec{\mathbf{r}}_{kl} + 2\vec{\mathbf{r}}_{ij}\cdot\vec{\mathbf{r}}_{jk} + 2\vec{\mathbf{r}}_{jk}\cdot\vec{\mathbf{r}}_{kl} + 2\vec{\mathbf{r}}_{ij}\cdot\vec{\mathbf{r}}_{kl}}. \quad (3.29)$$

With (3.7), (3.13), and (3.26), from (3.29) follows (3.28). $\square$

LEMMA 3.6

Let $\vec{\mathbf{e}}_a$ and $\vec{\mathbf{e}}_b$ be two non-collinear vectors, and let $\angle(\vec{\mathbf{e}}_a, \vec{\mathbf{e}}_b) =: \gamma$. Then there exist at maximum two different unit vectors $\vec{\mathbf{e}}_{c_i}$ $(i = \overline{1,2})$, such that

$$\angle(\vec{\mathbf{e}}_a, \vec{\mathbf{e}}_{c_i}) = \alpha \quad and \quad \angle(\vec{\mathbf{e}}_b, \vec{\mathbf{e}}_{c_i}) = \beta.$$

There is only one solution $\vec{\mathbf{e}}_{c_1}$, if vectors $\vec{\mathbf{e}}_a, \vec{\mathbf{e}}_b$ and $\vec{\mathbf{e}}_{c_1}$ are linearly dependent. Otherwise vectors $\vec{\mathbf{e}}_{c_1}$ and $\vec{\mathbf{e}}_{c_2}$ lay at the different sides of the plane P_{ab} spanned by $\vec{\mathbf{e}}_a$ and $\vec{\mathbf{e}}_b$.*

Proof. Let $\vec{\mathbf{e}}_c$ be a possible solution, i.e.

$$\|\vec{\mathbf{e}}_c\| = 1, \quad \angle(\vec{\mathbf{e}}_a, \vec{\mathbf{e}}_c) = \alpha \quad and \quad \angle(\vec{\mathbf{e}}_b, \vec{\mathbf{e}}_c) = \beta,$$

let

$$\vec{\mathbf{e}}_{ab}^{\perp} := \frac{\vec{\mathbf{e}}_a \times \vec{\mathbf{e}}_b}{\|\vec{\mathbf{e}}_a \times \vec{\mathbf{e}}_b\|},$$

and let

$$\vec{\mathbf{e}}_c = a\,\vec{\mathbf{e}}_a + b\,\vec{\mathbf{e}}_b + c\,\vec{\mathbf{e}}_{ab}^{\perp} \quad (3.30)$$

be the decomposition of $\vec{\mathbf{e}}_c$ in the basis $\vec{\mathbf{e}}_a, \vec{\mathbf{e}}_b, \vec{\mathbf{e}}_{ab}^{\perp}$. Then we have:

$$\vec{\mathbf{e}}_c \cdot \vec{\mathbf{e}}_a = a + b\cos\gamma \doteq \cos\alpha, \quad (3.31)$$
$$\vec{\mathbf{e}}_c \cdot \vec{\mathbf{e}}_b = a\cos\gamma + b \doteq \cos\beta. \quad (3.32)$$

We see that

$$\begin{vmatrix} 1 & \cos\gamma \\ \cos\gamma & 1 \end{vmatrix} = \sin^2\gamma \neq 0,$$

---

*Here vectors are considered relative to the same origin.

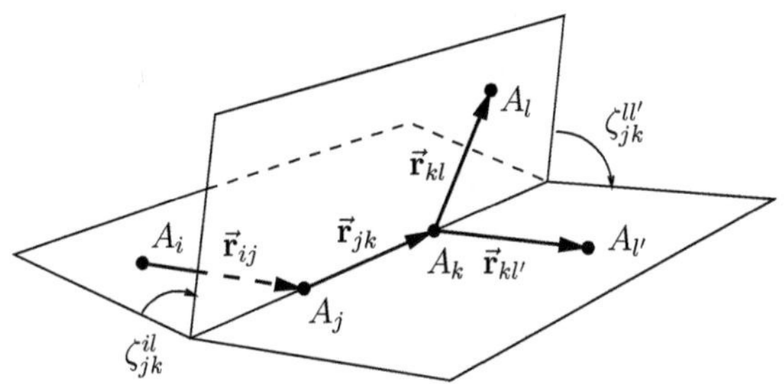

FIG. 3.1: The relation of dihedral angles in case of branching.

since $\vec{e}_a$ and $\vec{e}_b$ are non-collinear. Therefore system (3.31),(3.32) has a unique solution (a, b). Using the condition of the unit length, we can determine c:

$$c = \pm\sqrt{1 - a^2 - b^2}.$$

If $c \neq 0$, there are exactly two solutions:

$$\vec{e}_{c_1} = a\,\vec{e}_a + b\,\vec{e}_b + \sqrt{1 - a^2 - b^2}\,\vec{e}_{ab}^{\perp} \quad \text{end} \quad \vec{e}_{c_2} = a\,\vec{e}_a + b\,\vec{e}_b - \sqrt{1 - a^2 - b^2}\,\vec{e}_{ab}^{\perp}.$$

They lay at the different sides of the plane P_{ab}, since $\vec{e}_{c_1} \cdot \vec{e}_{ab}^{\perp} > 0$ and $\vec{e}_{c_2} \cdot \vec{e}_{ab}^{\perp} < 0$.

If $c = 0$, there is only one solution. It is clear from (3.30) that $c = 0$ iff $\vec{e}_a, \vec{e}_b$ and $\vec{e}_c$ are linearly dependent. $\square$

Proposition 3.7

Let A_i, A_j, A_k, A_l be the centers of four consequently connected atoms of a primary chain, $\alpha_k \in (0, \frac{2\pi}{3}]$, and let atom $A_{l'}$ be also bonded to A_k (Fig. 3.1). Then, for given $\vec{e}_{ij}, \vec{e}_{jk}$, and fixed bond angles, the direction $\vec{e}_{kl'}$ can be expressed as the following function of the primary dihedral angle ζ_{jk}^{il}:

$$\vec{e}_{kl'} = \vec{e}_{ijk}\sin\alpha_k\sin(\zeta_{jk}^{il} + \zeta_{jk}^{ll'}) - \vec{e}_{(i)jk}^{\perp}\sin\alpha_k\cos(\zeta_{jk}^{il} + \zeta_{jk}^{ll'}) - \vec{e}_{jk}\cos\alpha_k, \quad (3.33)$$

with $\vec{e}_{ijk}$ and $\vec{e}_{(i)jk}^{\perp}$ specified in (3.3) and $\zeta_{jk}^{ll'}$ given by:

$$\zeta_{jk}^{ll'} = \pm\arccos\left(\frac{\cos\alpha_k}{1 + \cos\alpha_k}\right). \quad (3.34)$$

The sign of $\zeta_{jk}^{ll'}$ depends on chiral configuration of the molecule and remains constant when the primary dihedral angle is changed. More precisely,

$$\text{sign}(\zeta_{jk}^{ll'}) = \text{sign}((\vec{r}_{jk} \times \vec{r}_{kl}) \cdot \vec{r}_{kl'}), \quad (3.35)$$

with the sign function defined in (1.43).

Proof. Obviously, for $\vec{\mathbf{e}}_{kl'}$ given by (3.33) holds:

$$\angle(\vec{\mathbf{e}}_{jk}, \vec{\mathbf{e}}_{kl'}) = \pi - \alpha_k, \tag{3.36}$$

since $\alpha_k \in (0, \pi)$ and

$$\vec{\mathbf{e}}_{jk} \cdot \vec{\mathbf{e}}_{kl'} = -\cos \alpha_k.$$

We show that

$$\angle(\vec{\mathbf{e}}_{kl}, \vec{\mathbf{e}}_{kl'}) = \alpha_k. \tag{3.37}$$

Utilizing (3.5) and (3.33) we deduce:

$$\vec{\mathbf{e}}_{kl} \cdot \vec{\mathbf{e}}_{kl'} = \sin^2 \alpha_k \sin \zeta_{jk}^{il} \sin(\zeta_{jk}^{il} + \zeta_{jk}^{ll'}) + \sin^2 \alpha_k \cos \zeta_{jk}^{il} \cos(\zeta_{jk}^{il} + \zeta_{jk}^{ll'}) + \cos^2 \alpha_k.$$

Further, using an addition theorem, we transform

$$\sin \zeta_{jk}^{il} \sin(\zeta_{jk}^{il} + \zeta_{jk}^{ll'}) + \cos \zeta_{jk}^{il} \cos(\zeta_{jk}^{il} + \zeta_{jk}^{ll'})$$

into

$$\cos(\zeta_{jk}^{il} + \zeta_{jk}^{ll'} - \zeta_{jk}^{il}) = \cos \zeta_{jk}^{ll'},$$

and with (3.34) we obtain:

$$\vec{\mathbf{e}}_{kl} \cdot \vec{\mathbf{e}}_{kl'} = \sin^2 \alpha_k \cos \zeta_{jk}^{ll'} + \cos^2 \alpha_k = \sin^2 \alpha_k \frac{\cos \alpha_k}{1 + \cos \alpha_k} + \cos^2 \alpha_k =$$

$$= \frac{\sin^2 \alpha_k \cos \alpha_k + \cos^2 \alpha_k(1 + \cos \alpha_k)}{1 + \cos \alpha_k} = \frac{\cos \alpha_k + \cos^2 \alpha_k}{1 + \cos \alpha_k} = \cos \alpha_k,$$

which implies validity of (3.37), since $\alpha_k \in (0, \pi)$.

To verify (3.35), we compute $(\vec{\mathbf{r}}_{jk} \times \vec{\mathbf{r}}_{kl}) \cdot \vec{\mathbf{r}}_{kl'}$ using (3.16), (3.17), (3.21), (3.33) and an addition theorem:

$$(\vec{\mathbf{r}}_{jk} \times \vec{\mathbf{r}}_{kl}) \cdot \vec{\mathbf{r}}_{kl'} = -\sin^2 \alpha_k \sin \zeta_{jk}^{il} \cos(\zeta_{jk}^{il} + \zeta_{jk}^{ll'}) + \sin^2 \alpha_k \cos \zeta_{jk}^{il} \sin(\zeta_{jk}^{il} + \zeta_{jk}^{ll'}) =$$

$$= \sin^2 \alpha_k \sin \zeta_{jk}^{ll'}.$$

Therefore, for $|\zeta_{jk}^{ll'}| \in (0, \pi)$ holds:

$$(\vec{\mathbf{r}}_{jk} \times \vec{\mathbf{r}}_{kl}) \cdot \vec{\mathbf{r}}_{kl'} < 0 \quad \text{if} \quad \zeta_{jk}^{ll'} < 0, \quad \text{and} \quad (\vec{\mathbf{r}}_{jk} \times \vec{\mathbf{r}}_{kl}) \cdot \vec{\mathbf{r}}_{kl'} > 0 \quad \text{if} \quad \zeta_{jk}^{ll'} > 0.$$

Thus, for $|\zeta_{jk}^{ll'}| \in (0, \pi)$, (3.33) equipped with (3.34) gives two solutions at the both sides* of plane P_{jkl} spanned by vectors $\vec{\mathbf{e}}_{jk}$ and $\vec{\mathbf{e}}_{kl}$.

If $|\zeta_{jk}^{ll'}| = \pi$, then $(\vec{\mathbf{r}}_{jk} \times \vec{\mathbf{r}}_{kl}) \cdot \vec{\mathbf{r}}_{kl'} = 0$, i.e. vector $\vec{\mathbf{e}}_{kl'}$ lays in plane P_{jkl}. Solution vectors for $\zeta_{jk}^{ll'} = \pi$ and $\zeta_{jk}^{ll'} = -\pi$ are identical. The corresponding $\alpha_k = \frac{2\pi}{3}$.

Consequently, by Lemma 3.6, there are no solutions, other then (3.33), that fulfill the conditions (3.36) and (3.37). $\square$

---

*Relative to A_k as an origin.

Definition 3.8

Let $\Theta : \mathbb{R} \to (-\pi, \pi]$ denote a surjection, such that

$$\Theta(\theta) := ((\theta - \pi) \bmod 2\pi) + \pi.$$

Remark 3.9

By convention, dihedral angles are defined in the interval $(-\pi, \pi]$. Some operations, such as angle addition, may cause violations of the boundaries of the interval, still giving a physically meaningful answer. The function of Θ is to map an angle to a formally equivalent one in the desired interval.

Corollary 3.10

Under the conditions of Proposition 3.7, the dihedral angle $\angle(\vec{r}_{ij}, \vec{r}_{jk}, \vec{r}_{kl'})$ can be expressed as a sum of the corresponding primary dihedral angle and a supplement with a constant magnitude depending only on a related bond angle:

$$\zeta_{jk}^{il'} = \Theta(\zeta_{jk}^{il} + \zeta_{jk}^{ll'}), \tag{3.38}$$

where $\zeta_{jk}^{ll'}$ is given by:

$$\zeta_{jk}^{ll'} = \mathrm{sign}((\vec{r}_{jk} \times \vec{r}_{kl}) \cdot \vec{r}_{kl'}) \arccos\left(\frac{\cos \alpha_k}{1 + \cos \alpha_k}\right) \tag{3.39}$$

with the sign function defined in (1.43).

Proof. Follows from Propositions 3.3 and 3.7. $\square$

Corollary 3.11

Let A_i, A_j, A_k, A_l be the centers of four consequently connected atoms of a primary chain, and let atoms $A_{i'}$ and $A_{l'}$ be bonded to A_j and A_k respectively. Then the dihedral angles $\angle(\vec{r}_{i'j}, \vec{r}_{jk}, \vec{r}_{kl})$ and $\angle(\vec{r}_{i'j}, \vec{r}_{jk}, \vec{r}_{kl'}))$ are given by:

$$\zeta_{jk}^{i'l} = \Theta(\zeta_{jk}^{il} + \zeta_{jk}^{ii'}) \tag{3.40}$$

and

$$\zeta_{jk}^{i'l'} = \Theta(\zeta_{jk}^{il} + \zeta_{jk}^{ii'} + \zeta_{jk}^{ll'}), \tag{3.41}$$

with

$$\zeta_{jk}^{ii'} = \mathrm{sign}((\vec{r}_{ij} \times \vec{r}_{jk}) \cdot \vec{r}_{i'j}) \arccos\left(\frac{\cos \alpha_j}{1 + \cos \alpha_j}\right) \tag{3.42}$$

and $\zeta_{jk}^{ll'}$ defined in (3.39).

Proof. First of all, note that

$$\zeta_{kj}^{li} = \zeta_{jk}^{il}.$$

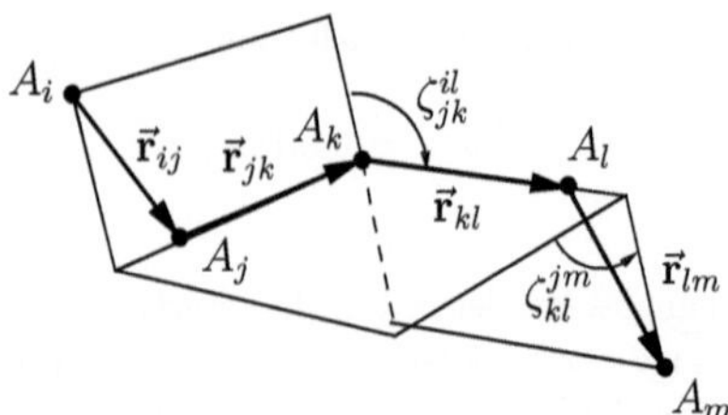

FIG. 3.2: Centers A_i, A_j, A_k, A_l, A_m of five consequently bonded atoms, and the related dihedral angles.

Indeed,

$$\zeta_{kj}^{li} = \text{sign}((\vec{r}_{kj} \times \vec{r}_{ji}) \cdot \vec{r}_{lk}) \arccos\left(\frac{(\vec{r}_{lk} \times \vec{r}_{kj}) \cdot (\vec{r}_{kj} \times \vec{r}_{ji})}{\|\vec{r}_{lk} \times \vec{r}_{kj}\| \|\vec{r}_{kj} \times \vec{r}_{ji}\|}\right) =$$

$$= \text{sign}(((-\vec{r}_{jk}) \times (-\vec{r}_{ij})) \cdot (-\vec{r}_{kl})) \arccos\left(\frac{((-\vec{r}_{kl}) \times (-\vec{r}_{jk})) \cdot ((-\vec{r}_{jk}) \times (-\vec{r}_{ij}))}{\|(-\vec{r}_{kl}) \times (-\vec{r}_{jk})\| \|(-\vec{r}_{jk}) \times (-\vec{r}_{ij})\|}\right) =$$

$$= \text{sign}((-\vec{r}_{ij} \times \vec{r}_{jk}) \cdot (-\vec{r}_{kl})) \arccos\left(\frac{(-\vec{r}_{jk} \times \vec{r}_{kl}) \cdot (-\vec{r}_{ij} \times \vec{r}_{jk})}{\|\vec{r}_{jk} \times \vec{r}_{kl}\| \|\vec{r}_{ij} \times \vec{r}_{jk}\|}\right) =$$

$$= \text{sign}((\vec{r}_{jk} \times \vec{r}_{kl}) \cdot \vec{r}_{ij}) \arccos\left(\frac{(\vec{r}_{jk} \times \vec{r}_{kl}) \cdot (\vec{r}_{ij} \times \vec{r}_{jk})}{\|\vec{r}_{jk} \times \vec{r}_{kl}\| \|\vec{r}_{ij} \times \vec{r}_{jk}\|}\right) = \zeta_{jk}^{il}. \quad (3.43)$$

Similarly,

$$\zeta_{jk}^{i'l} = \zeta_{kj}^{li'}, \quad (3.44)$$

and with Corollary 3.10 follows:

$$\zeta_{kj}^{li'} = \Theta\left(\zeta_{kj}^{li} + \text{sign}((\vec{r}_{kj} \times \vec{r}_{ji}) \cdot \vec{r}_{ji'}) \arccos\left(\frac{\cos\alpha_j}{1 + \cos\alpha_j}\right)\right) =$$

$$= \Theta\left(\zeta_{jk}^{il} + \text{sign}((-(-\vec{r}_{ij}) \times (-\vec{r}_{jk})) \cdot (-\vec{r}_{i'j})) \arccos\left(\frac{\cos\alpha_j}{1 + \cos\alpha_j}\right)\right) =$$

$$= \Theta\left(\zeta_{jk}^{il} + \text{sign}((\vec{r}_{ij} \times \vec{r}_{jk}) \cdot \vec{r}_{i'j}) \arccos\left(\frac{\cos\alpha_j}{1 + \cos\alpha_j}\right)\right). \quad (3.45)$$

From (3.44) and (3.45) we immediately obtain (3.40).

Additionally, we see that

$$\zeta_{jk}^{i'l'} = \Theta(\zeta_{jk}^{i'l} + \zeta_{jk}^{ll'}),$$

hence with (3.40) follows (3.41). □

PROPOSITION 3.12
Let A_i, A_j, A_k, A_l and A_m be the centers of five consequently bonded atoms

(Fig. 3.2). Then the distance between A_i and A_l is given by:

$$\|\vec{\mathbf{r}}_{im}\| = \sqrt{\|\vec{\mathbf{r}}_{il}\|^2 + l_{lm}^2 - 2l_{kl}l_{lm}\cos\alpha_l + 2l_{jk}l_{lm}(\cos\alpha_k\cos\alpha_l - \sin\alpha_k\sin\alpha_l\cos\zeta_{kl}^{jm})+}$$

$$+2l_{ij}l_{lm}(\sin\alpha_k(\sin\alpha_j\cos\alpha_l\cos\zeta_{jk}^{il} + \cos\alpha_j\sin\alpha_l\cos\zeta_{kl}^{jm})+$$

$$+ \sin\alpha_j\sin\alpha_l(\sin\zeta_{jk}^{il}\sin\zeta_{kl}^{jm} + \cos\alpha_k\cos\zeta_{jk}^{il}\cos\zeta_{kl}^{jm}) - \cos\alpha_j\cos\alpha_k\cos\alpha_l). \quad (3.46)$$

Proof. First of all, we note that

$$\|\vec{\mathbf{r}}_{im}\| = \sqrt{(\vec{\mathbf{r}}_{il} + \vec{\mathbf{r}}_{lm})\cdot(\vec{\mathbf{r}}_{il} + \vec{\mathbf{r}}_{lm})} = \sqrt{\|\vec{\mathbf{r}}_{il}\|^2 + l_{lm}^2 + 2(\vec{\mathbf{r}}_{ij} + \vec{\mathbf{r}}_{jk} + \vec{\mathbf{r}}_{kl})\cdot\vec{\mathbf{r}}_{lm}} =$$

$$= \sqrt{\|\vec{\mathbf{r}}_{il}\|^2 + l_{lm}^2 + 2l_{ij}l_{lm}\vec{\mathbf{e}}_{ij}\cdot\vec{\mathbf{e}}_{lm} + 2l_{jk}l_{lm}\vec{\mathbf{e}}_{jk}\cdot\vec{\mathbf{e}}_{lm} + 2l_{kl}l_{lm}\vec{\mathbf{e}}_{kl}\cdot\vec{\mathbf{e}}_{lm}}. \quad (3.47)$$

To compute $\vec{\mathbf{e}}_{ij}\cdot\vec{\mathbf{e}}_{lm}$, we express $\vec{\mathbf{e}}_{ij}$ and $\vec{\mathbf{e}}_{lm}$ in the basis

$$\vec{\mathbf{e}}_{jk}, \quad \vec{\mathbf{e}}_{jkl} := \frac{\vec{\mathbf{e}}_{jk}\times\vec{\mathbf{e}}_{kl}}{\|\vec{\mathbf{e}}_{jk}\times\vec{\mathbf{e}}_{kl}\|}, \quad \vec{\mathbf{e}}_{jk(l)}^{\perp} := \vec{\mathbf{e}}_{jk}\times\vec{\mathbf{e}}_{jkl}. \quad (3.48)$$

Exchanging the roles of indexes i, j, k, l in (3.5) in such a way, that they are considered in the opposite order, we obtain:

$$\vec{\mathbf{e}}_{ji} = \vec{\mathbf{e}}_{lkj}\sin\alpha_j\sin\zeta_{kj}^{li} - \vec{\mathbf{e}}_{(l)kj}^{\perp}\sin\alpha_j\cos\zeta_{kj}^{li} - \vec{\mathbf{e}}_{kj}\cos\alpha_j.$$

We note that

$$\vec{\mathbf{e}}_{ji} = -\vec{\mathbf{e}}_{ij}, \ \vec{\mathbf{e}}_{lkj} = \frac{\vec{\mathbf{e}}_{lk}\times\vec{\mathbf{e}}_{kj}}{\sin\alpha_k} = \frac{(-\vec{\mathbf{e}}_{kl})\times(-\vec{\mathbf{e}}_{jk})}{\sin\alpha_k} = -\frac{\vec{\mathbf{e}}_{jk}\times\vec{\mathbf{e}}_{kl}}{\sin\alpha_k} = -\vec{\mathbf{e}}_{jkl},$$

$$\vec{\mathbf{e}}_{(l)kj}^{\perp} = \vec{\mathbf{e}}_{kj}\times\vec{\mathbf{e}}_{lkj} = (-\vec{\mathbf{e}}_{jk})\times(-\vec{\mathbf{e}}_{jkl}) = \vec{\mathbf{e}}_{jk(l)}^{\perp}, \ \vec{\mathbf{e}}_{kj} = -\vec{\mathbf{e}}_{jk}, \ \zeta_{kj}^{li} = \zeta_{jk}^{il},$$

hence

$$\vec{\mathbf{e}}_{ij} = \vec{\mathbf{e}}_{jkl}\sin\alpha_j\sin\zeta_{jk}^{il} + \vec{\mathbf{e}}_{jk(l)}^{\perp}\sin\alpha_j\cos\zeta_{jk}^{il} - \vec{\mathbf{e}}_{jk}\cos\alpha_j. \quad (3.49)$$

Similarly we can express $\vec{\mathbf{e}}_{lm}$ in the basis (3.48) using (3.6):

$$\vec{\mathbf{e}}_{lm} = \vec{\mathbf{e}}_{jkl}\sin\alpha_l\sin\zeta_{kl}^{jm} + \vec{\mathbf{e}}_{jk(l)}^{\perp}(\sin\alpha_k\cos\alpha_l + \cos\alpha_k\sin\alpha_l\cos\zeta_{kl}^{jm})+$$

$$+ \vec{\mathbf{e}}_{jk}(\cos\alpha_k\cos\alpha_l - \sin\alpha_k\sin\alpha_l\cos\zeta_{kl}^{jm}). \quad (3.50)$$

From (3.49) and (3.50) we deduce:

$$\vec{\mathbf{e}}_{ij}\cdot\vec{\mathbf{e}}_{lm} = \sin\alpha_k(\sin\alpha_j\cos\alpha_l\cos\zeta_{jk}^{il} + \cos\alpha_j\sin\alpha_l\cos\zeta_{kl}^{jm})+$$

$$+ \sin\alpha_j\sin\alpha_l(\sin\zeta_{jk}^{il}\sin\zeta_{kl}^{jm} + \cos\alpha_k\cos\zeta_{jk}^{il}\cos\zeta_{kl}^{jm})-$$

$$- \cos\alpha_j\cos\alpha_k\cos\alpha_l. \quad (3.51)$$

Shifting indexes in (3.26) gives:

$$\vec{\mathbf{e}}_{jk}\cdot\vec{\mathbf{e}}_{lm} = \cos\alpha_k\cos\alpha_l - \sin\alpha_k\sin\alpha_l\cos\zeta_{kl}^{jm}, \quad (3.52)$$

and similar to (3.7) holds:

$$\vec{\mathbf{e}}_{kl}\cdot\vec{\mathbf{e}}_{lm} = -\cos\alpha_l. \quad (3.53)$$

Substitution of (3.51), (3.52), and (3.53) into (3.47) gives (3.46). $\square$

Definition 3.13

Let $\Xi_\mathcal{B} : \mathbb{R}^3 \to \mathbb{R}^3$ denote a linear map such that

$$\Xi_\mathcal{B}(\vec{\mathbf{v}}) := \begin{pmatrix} \vec{\mathbf{a}}^{\mathrm{T}} \\ \vec{\mathbf{b}}^{\mathrm{T}} \\ \vec{\mathbf{c}}^{\mathrm{T}} \end{pmatrix} \vec{\mathbf{v}}, \quad \forall\, \vec{\mathbf{v}} \in \mathbb{R}^3,$$

where $\mathcal{B} := \{\vec{\mathbf{a}}, \vec{\mathbf{b}}, \vec{\mathbf{c}}\}$ is an ordered orthonormal basis.

Lemma 3.14

$\Xi_\mathcal{B}$ associates a vector $\vec{\mathbf{v}} \in \mathbb{R}^3$ with its coordinate representation in the basis $\mathcal{B}$, i.e. for $\mathcal{B} =: \{\vec{\mathbf{a}}, \vec{\mathbf{b}}, \vec{\mathbf{c}}\}$ and

$$\begin{pmatrix} w^{[1]} \\ w^{[2]} \\ w^{[3]} \end{pmatrix} := \vec{\mathbf{w}} := \Xi_\mathcal{B}(\vec{\mathbf{v}})$$

holds:

$$w^{[1]}\vec{\mathbf{a}} + w^{[2]}\vec{\mathbf{b}} + w^{[3]}\vec{\mathbf{c}} = \vec{\mathbf{v}}. \tag{3.54}$$

Proof. Let

$$\vec{\mathbf{v}} = a\,\vec{\mathbf{a}} + b\,\vec{\mathbf{b}} + c\,\vec{\mathbf{c}}.$$

Then we have:

$$\Xi_\mathcal{B}(\vec{\mathbf{v}}) = \begin{pmatrix} \vec{\mathbf{a}}^{\mathrm{T}} \\ \vec{\mathbf{b}}^{\mathrm{T}} \\ \vec{\mathbf{c}}^{\mathrm{T}} \end{pmatrix} (a\,\vec{\mathbf{a}} + b\,\vec{\mathbf{b}} + c\,\vec{\mathbf{c}}) = \begin{pmatrix} \vec{\mathbf{a}} \cdot (a\,\vec{\mathbf{a}} + b\,\vec{\mathbf{b}} + c\,\vec{\mathbf{c}}) \\ \vec{\mathbf{b}} \cdot (a\,\vec{\mathbf{a}} + b\,\vec{\mathbf{b}} + c\,\vec{\mathbf{c}}) \\ \vec{\mathbf{c}} \cdot (a\,\vec{\mathbf{a}} + b\,\vec{\mathbf{b}} + c\,\vec{\mathbf{c}}) \end{pmatrix} = \begin{pmatrix} a \\ b \\ c \end{pmatrix} . \quad \square$$

Remark 3.15

If $\mathcal{B}_a := \{\vec{\mathbf{e}}_1^a, \vec{\mathbf{e}}_2^a, \vec{\mathbf{e}}_3^a\}$ and $\mathcal{B}_b := \{\vec{\mathbf{e}}_1^b, \vec{\mathbf{e}}_2^b, \vec{\mathbf{e}}_3^b\}$ are two ordered orthonormal bases in $\mathbb{R}^3$, $\vec{\mathbf{v}} \in \mathbb{R}^3$, and $\vec{\mathbf{b}} := \Xi_{\mathcal{B}_b}(\vec{\mathbf{v}})$, then $\Xi_{\mathcal{B}_a} \circ \Xi_{\mathcal{B}_b}^{-1}(\vec{\mathbf{b}})$ maps the coordinate representation of $\vec{\mathbf{v}}$ in $\mathcal{B}_b$ into its coordinate representation in $\mathcal{B}_a$, and the following holds:

$$\Xi_{\mathcal{B}_a} \circ \Xi_{\mathcal{B}_b}^{-1}(\vec{\mathbf{b}}) = \begin{pmatrix} (\vec{\mathbf{e}}_1^a)^{\mathrm{T}} \\ (\vec{\mathbf{e}}_2^a)^{\mathrm{T}} \\ (\vec{\mathbf{e}}_3^a)^{\mathrm{T}} \end{pmatrix} \begin{pmatrix} (\vec{\mathbf{e}}_1^b)^{\mathrm{T}} \\ (\vec{\mathbf{e}}_2^b)^{\mathrm{T}} \\ (\vec{\mathbf{e}}_3^b)^{\mathrm{T}} \end{pmatrix}^{-1} \vec{\mathbf{b}} = \begin{pmatrix} (\vec{\mathbf{e}}_1^a)^{\mathrm{T}} \\ (\vec{\mathbf{e}}_2^a)^{\mathrm{T}} \\ (\vec{\mathbf{e}}_3^a)^{\mathrm{T}} \end{pmatrix} \begin{pmatrix} (\vec{\mathbf{e}}_1^b)^{\mathrm{T}} \\ (\vec{\mathbf{e}}_2^b)^{\mathrm{T}} \\ (\vec{\mathbf{e}}_3^b)^{\mathrm{T}} \end{pmatrix}^{\mathrm{T}} \vec{\mathbf{b}} =$$

$$= \begin{pmatrix} (\vec{\mathbf{e}}_1^a)^{\mathrm{T}} \\ (\vec{\mathbf{e}}_2^a)^{\mathrm{T}} \\ (\vec{\mathbf{e}}_3^a)^{\mathrm{T}} \end{pmatrix} \begin{pmatrix} \vec{\mathbf{e}}_1^b & \vec{\mathbf{e}}_2^b & \vec{\mathbf{e}}_1^b \end{pmatrix} \vec{\mathbf{b}} = \begin{pmatrix} \vec{\mathbf{e}}_1^a \cdot \vec{\mathbf{e}}_1^b & \vec{\mathbf{e}}_1^a \cdot \vec{\mathbf{e}}_2^b & \vec{\mathbf{e}}_1^a \cdot \vec{\mathbf{e}}_3^b \\ \vec{\mathbf{e}}_2^a \cdot \vec{\mathbf{e}}_1^b & \vec{\mathbf{e}}_2^a \cdot \vec{\mathbf{e}}_2^b & \vec{\mathbf{e}}_2^a \cdot \vec{\mathbf{e}}_3^b \\ \vec{\mathbf{e}}_3^a \cdot \vec{\mathbf{e}}_1^b & \vec{\mathbf{e}}_3^a \cdot \vec{\mathbf{e}}_2^b & \vec{\mathbf{e}}_3^a \cdot \vec{\mathbf{e}}_3^b \end{pmatrix} \vec{\mathbf{b}}.$$

Besides, if $\vec{w} \in \mathbb{R}^3$ and $\vec{a} := \Xi_{\mathcal{B}_a}(\vec{w})$, then

$$\vec{w} \cdot \vec{v} = \Xi_{\mathcal{B}_a}^{-1}(\vec{a}) \cdot \Xi_{\mathcal{B}_b}^{-1}(\vec{b}) = \left(\begin{pmatrix} (\vec{e}_1^{\,a})^{\mathrm{T}} \\ (\vec{e}_2^{\,a})^{\mathrm{T}} \\ (\vec{e}_3^{\,a})^{\mathrm{T}} \end{pmatrix}^{-1} \vec{a} \right)^{\mathrm{T}} \begin{pmatrix} (\vec{e}_1^{\,b})^{\mathrm{T}} \\ (\vec{e}_2^{\,b})^{\mathrm{T}} \\ (\vec{e}_3^{\,b})^{\mathrm{T}} \end{pmatrix}^{-1} \vec{b} =$$

$$= \vec{a}^{\mathrm{T}} \begin{pmatrix} (\vec{e}_1^{\,a})^{\mathrm{T}} \\ (\vec{e}_2^{\,a})^{\mathrm{T}} \\ (\vec{e}_3^{\,a})^{\mathrm{T}} \end{pmatrix} \begin{pmatrix} (\vec{e}_1^{\,b})^{\mathrm{T}} \\ (\vec{e}_2^{\,b})^{\mathrm{T}} \\ (\vec{e}_3^{\,b})^{\mathrm{T}} \end{pmatrix}^{-1} \vec{b} = \vec{a}^{\mathrm{T}} \, \Xi_{\mathcal{B}_a} \circ \Xi_{\mathcal{B}_b}^{-1}(\vec{b}).$$

LEMMA 3.16

Let A_i, A_j, A_k and A_l be the centers of four consequently bonded atoms, and let $\mathcal{B}_{ijk}$ and $\mathcal{B}_{jkl}$ refer to the bases (3.4) and (3.48) respectively. Then the mapping $\Xi_{\mathcal{B}_{ijk}} \circ \Xi_{\mathcal{B}_{jkl}}^{-1}$ is given by

$$\Xi_{\mathcal{B}_{ijk}} \circ \Xi_{\mathcal{B}_{jkl}}^{-1}(\vec{v}) = \mathbf{A}_{jk}^{il} \vec{v}, \quad \forall \vec{v} \in \mathbb{R}^3,$$

where

$$\mathbf{A}_{jk}^{il} = \begin{pmatrix} -\cos\alpha_j & \sin\alpha_j \sin\zeta_{jk}^{il} & \sin\alpha_j \cos\zeta_{jk}^{il} \\ 0 & \cos\zeta_{jk}^{il} & -\sin\zeta_{jk}^{il} \\ -\sin\alpha_j & -\cos\alpha_j \sin\zeta_{jk}^{il} & -\cos\alpha_j \cos\zeta_{jk}^{il} \end{pmatrix}. \tag{3.55}$$

Proof. From Remark 3.15 follows that

$$\mathbf{A}_{jk}^{il} = \begin{pmatrix} \vec{e}_{ij} \cdot \vec{e}_{jk} & \vec{e}_{ij} \cdot \vec{e}_{jkl} & \vec{e}_{ij} \cdot \vec{e}_{jk(l)}^{\perp} \\ \vec{e}_{ijk} \cdot \vec{e}_{jk} & \vec{e}_{ijk} \cdot \vec{e}_{jkl} & \vec{e}_{ijk} \cdot \vec{e}_{jk(l)}^{\perp} \\ \vec{e}_{ij(k)}^{\perp} \cdot \vec{e}_{jk} & \vec{e}_{ij(k)}^{\perp} \cdot \vec{e}_{jkl} & \vec{e}_{ij(k)}^{\perp} \cdot \vec{e}_{jk(l)}^{\perp} \end{pmatrix} \tag{3.56}$$

In accordance with (1.46), we have:

$$\vec{e}_{ijk} \cdot \vec{e}_{jkl} = \cos\zeta_{jk}^{il}, \tag{3.57}$$

hence

$$\vec{e}_{ij} \cdot \vec{e}_{jk(l)}^{\perp} = \vec{e}_{ij} \cdot (\vec{e}_{jk} \times \vec{e}_{jkl}) = (\vec{e}_{ij} \times \vec{e}_{jk}) \cdot \vec{e}_{jkl} =$$
$$= \sin\alpha_j \, \vec{e}_{ijk} \cdot \vec{e}_{jkl} = \sin\alpha_j \cos\zeta_{jk}^{il}, \quad (3.58)$$

and with (3.23) follows:

$$\vec{e}_{ij} \cdot \vec{e}_{jkl} = \frac{\vec{e}_{ij} \cdot (\vec{e}_{jk} \times \vec{e}_{kl})}{\sin\alpha_k} = \frac{(\vec{e}_{ij} \times \vec{e}_{jk}) \cdot \vec{e}_{kl}}{\sin\alpha_k} =$$
$$= \frac{\sin\alpha_j \, \vec{e}_{ijk} \cdot \vec{e}_{kl}}{\sin\alpha_k} = \sin\alpha_j \sin\zeta_{jk}^{il}. \quad (3.59)$$

Using Lagrange's identity from vector calculus, (3.7), (3.57), (3.23), and (3.59) we deduce:

$$\vec{\mathbf{e}}^{\perp}_{ij(k)} \cdot \vec{\mathbf{e}}^{\perp}_{jk(l)} = (\vec{\mathbf{e}}_{ij} \times \vec{\mathbf{e}}_{ijk}) \cdot (\vec{\mathbf{e}}_{jk} \times \vec{\mathbf{e}}_{jkl}) =$$
$$= (\vec{\mathbf{e}}_{ij} \cdot \vec{\mathbf{e}}_{jk})(\vec{\mathbf{e}}_{ijk} \cdot \vec{\mathbf{e}}_{jkl}) - (\vec{\mathbf{e}}_{ijk} \cdot \vec{\mathbf{e}}_{jk})(\vec{\mathbf{e}}_{ij} \cdot \vec{\mathbf{e}}_{jkl}) = -\cos \alpha_j \cos \zeta^{il}_{jk}, \quad (3.60)$$

$$\vec{\mathbf{e}}^{\perp}_{ij(k)} \cdot \vec{\mathbf{e}}_{jkl} = (\vec{\mathbf{e}}_{ij} \times \vec{\mathbf{e}}_{ijk}) \cdot \frac{\vec{\mathbf{e}}_{jk} \times \vec{\mathbf{e}}_{kl}}{\sin \alpha_k} =$$
$$= \frac{1}{\sin \alpha_k}((\vec{\mathbf{e}}_{ij} \cdot \vec{\mathbf{e}}_{jk})(\vec{\mathbf{e}}_{ijk} \cdot \vec{\mathbf{e}}_{kl}) - (\vec{\mathbf{e}}_{ij} \cdot \vec{\mathbf{e}}_{kl})(\vec{\mathbf{e}}_{ijk} \cdot \vec{\mathbf{e}}_{jk})) = -\cos \alpha_j \sin \zeta^{il}_{jk}, \quad (3.61)$$

$$\vec{\mathbf{e}}_{ijk} \cdot \vec{\mathbf{e}}^{\perp}_{jk(l)} = \frac{\vec{\mathbf{e}}_{ij} \times \vec{\mathbf{e}}_{jk}}{\sin \alpha_j} \cdot (\vec{\mathbf{e}}_{jk} \times \vec{\mathbf{e}}_{jkl}) =$$
$$= \frac{1}{\sin \alpha_j}((\vec{\mathbf{e}}_{ij} \cdot \vec{\mathbf{e}}_{jk})(\vec{\mathbf{e}}_{jk} \cdot \vec{\mathbf{e}}_{jkl}) - (\vec{\mathbf{e}}_{ij} \cdot \vec{\mathbf{e}}_{jkl})(\vec{\mathbf{e}}_{jk} \cdot \vec{\mathbf{e}}_{jk})) = -\sin \zeta^{il}_{jk}. \quad (3.62)$$

Finally, substituting (3.7), (3.24), and (3.57)-(3.62) into (3.56) we obtain (3.55). $\square$

THEOREM 3.17

Let A_i, A_{i+1}, $\ldots$, A_{i+n} be the centers of consequently bonded atoms. Then the distance between A_i and A_{i+n} is given by:

$$\|\vec{\mathbf{r}}_{i\,i+n}\| = \sqrt{\sum_{j=i+1}^{i+n} l^2_{j-1\,j} + 2 \sum_{j=i+1}^{i+n-1} \sum_{k=j}^{i+n-1} l_{j-1\,j} l_{k\,k+1} \left(\prod_{m=j}^{k} \mathbf{A}^{m-1\,m+2}_{m\,m+1} \right)_{11}} =$$

$$= \Bigg(\sum_{j=i+1}^{i+n} l^2_{j-1\,j} + 2 \sum_{j=i+1}^{i+n-1} \sum_{k=j}^{i+n-1} l_{j-1\,j} l_{k\,k+1} \sum_{i_j=1}^{3} \sum_{i_{j+1}=1}^{3} \cdots \sum_{i_{k-2}=1}^{3} \sum_{i_{k-1}=1}^{3} (\mathbf{A}^{j-1\,j+2}_{j\,j+1})_{1 i_j}$$

$$(\mathbf{A}^{j\,j+3}_{j+1\,j+2})_{i_j i_{j+1}} \cdots (\mathbf{A}^{k-2\,k+1}_{k-1\,k})_{i_{k-2} i_{k-1}} (\mathbf{A}^{k-1\,k+2}_{k\,k+1})_{i_{k-1} 1} \Bigg)^{1/2}, \quad (3.63)$$

where

$$\mathbf{A}^{m-1\,m+2}_{m\,m+1} = \begin{pmatrix} -\cos \alpha_m & \sin \alpha_m \sin \zeta^{m-1\,m+2}_{m\,m+1} & \sin \alpha_m \cos \zeta^{m-1\,m+2}_{m\,m+1} \\ 0 & \cos \zeta^{m-1\,m+2}_{m\,m+1} & -\sin \zeta^{m-1\,m+2}_{m\,m+1} \\ -\sin \alpha_m & -\cos \alpha_m \sin \zeta^{m-1\,m+2}_{m\,m+1} & -\cos \alpha_m \cos \zeta^{m-1\,m+2}_{m\,m+1} \end{pmatrix} \quad (3.64)$$

for

$$m = \overline{i+1, i+n-2},$$

and

$$\mathbf{A}^{i+n-2\,i+n+1*}_{i+n-1\,i+n} = \begin{pmatrix} -\cos \alpha_{i+n-1} & 0 & 0 \\ 0 & 0 & 0 \\ -\sin \alpha_{i+n-1} & 0 & 0 \end{pmatrix}. \quad (3.65)$$

---

*The second and the third column of $\mathbf{A}^{i+n-2\,i+n+1}_{i+n-1\,i+n}$ is of no relevance, therefore atom A_{i+n+1}, as well as the dihedral angle $\zeta^{i+n-2\,i+n+1}_{i+n-1\,i+n}$, does not have to exist.

Proof. First of all, note that

$$\|\vec{\mathbf{r}}_{i\,i+n}\| = \sqrt{(\sum_{j=i+1}^{i+n}\vec{\mathbf{r}}_{j-1\,j})(\sum_{j=i+1}^{i+n}\vec{\mathbf{r}}_{j-1\,j})} =$$

$$= \sqrt{\sum_{j=i+1}^{i+n} l_{j-1\,j}^2 + 2\sum_{j=i+1}^{i+n-1}\sum_{k=j}^{i+n-1} l_{j-1\,j}l_{k\,k+1}\vec{\mathbf{e}}_{j-1\,j}\cdot\vec{\mathbf{e}}_{k\,k+1}.} \tag{3.66}$$

Using Remark 3.15 and Lemma 3.16, we deduce by induction:

$$\vec{\mathbf{e}}_{j-1\,j}\cdot\vec{\mathbf{e}}_{k\,k+1} = \begin{pmatrix}1 & 0 & 0\end{pmatrix} \mathbf{A}_{j\,j+1}^{j-1\,j+2}\mathbf{A}_{j+1\,j+2}^{j\,j+3}\cdots\mathbf{A}_{k\,k+1}^{k-1\,k+2}\begin{pmatrix}1\\0\\0\end{pmatrix} =$$

$$= \left(\prod_{m=j}^{k}\mathbf{A}_{m\,m+1}^{m-1\,m+2}\right)_{11} = \sum_{i_j=1}^{3}\sum_{i_{j+1}=1}^{3}\cdots\sum_{i_{k-2}=1}^{3}\sum_{i_{k-1}=1}^{3}(\mathbf{A}_{j\,j+1}^{j-1\,j+2})_{1i_j}(\mathbf{A}_{j+1\,j+2}^{j\,j+3})_{i_j i_{j+1}}\cdots$$

$$\cdots(\mathbf{A}_{k-1\,k}^{k-2\,k+1})_{i_{k-2}i_{k-1}}(\mathbf{A}_{k\,k+1}^{k-1\,k+2})_{i_{k-1}1}, \tag{3.67}$$

where $\mathbf{A}_{i+1\,i+2}^{i\,i+3},\ldots,\mathbf{A}_{i+n-1\,i+n}^{i+n-2\,i+n+1}$ are given by (3.55) with respectively changed index notations (cf. (3.64)-(3.65)). Substitution of (3.67) into (3.66) gives (3.63). $\square$

COROLLARY 3.18
In a chain of consequently bonded atoms A_i, A_{i+1}, $\ldots$, A_{i+n} the distance between A_i and A_{i+n} is given by:

$$\|\vec{\mathbf{r}}_{i\,i+n}\| = \left[\|\vec{\mathbf{r}}_{i\,i+n-1}\|^2 + l_{i+n-1\,i+n}^2 + 2l_{i+n-1\,i+n}\sum_{j=i+1}^{i+n-1} l_{j-1\,j}\sum_{k_j=1}^{3}\sum_{k_{j+1}=1}^{3}\cdots\right.$$

$$\cdots\sum_{k_{i+n-3}=1}^{3}\sum_{k_{i+n-2}=1}^{3}(\mathbf{A}_{j\,j+1}^{j-1\,j+2})_{1k_j}(\mathbf{A}_{j+1\,j+2}^{j\,j+3})_{k_j k_{j+1}}\cdots$$

$$\left.\cdots(\mathbf{A}_{i+n-2\,i+n-1}^{i+n-3\,i+n})_{k_{i+n-3}k_{i+n-2}}(\mathbf{A}_{i+n-1\,i+n}^{i+n-2\,i+n+1})_{k_{i+n-2}1}\right., \tag{3.68}$$

where $\mathbf{A}_{i+1\,i+2}^{i\,i+3},\ldots,\mathbf{A}_{i+n-1\,i+n}^{i+n-2\,i+n+1}$ are defined as in Theorem 3.17.

Certain regularities in protein structure permit further simplifications. As justified in Subsection 1.5.2, sp^2-hybridized atoms, such as N and C_α, which participate in peptide bond formation, have the equilibrium bond angle $\frac{2\pi}{3}$. Using the well known relations

$$\cos\frac{\pi}{3} = \frac{1}{2} \quad \text{and} \quad \sin\frac{\pi}{3} = \frac{\sqrt{3}}{2},$$

we easily deduce:

$$\cos\frac{2\pi}{3} = \cos\left(\pi - \tfrac{\pi}{3}\right) = -\frac{1}{2} \quad \text{and} \quad \sin\frac{2\pi}{3} = \sin\left(\pi - \tfrac{\pi}{3}\right) = \frac{\sqrt{3}}{2}. \tag{3.69}$$

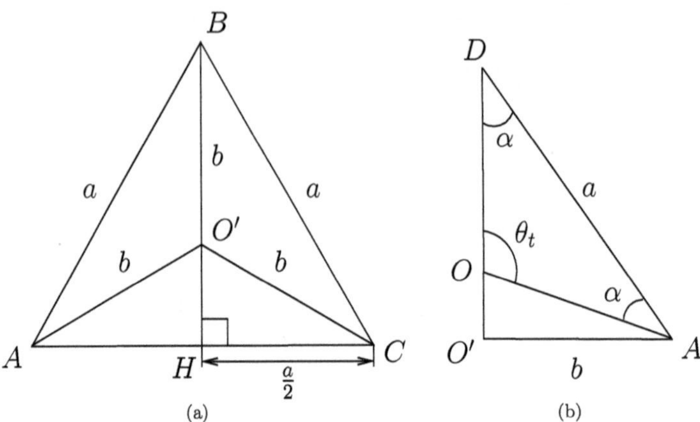

FIG. 3.3: (a) A face of a regular tetrahedron. $AB = BC = AC =: a$, $AO' = BO' = CO' =: b$, $\angle AO'B = \angle BO'C = \angle AO'C = \frac{2\pi}{3}$. (b) A cross-section through two vertices, A and D, and the center O of the tetrahedron. $AD = a$, $AO' = b$, $\angle ODA = \angle OAD =: \alpha$, $\vartheta_t := \angle AOD$.

The values of sine and cosine of the equilibrium angle of sp^3-hybridized atoms, for example α-carbons and the most of the side chain non-hydrogenic atoms, are given by the following lemma.

LEMMA 3.19

Let ϑ_t be the angular distance between two vertices of a regular tetrahedron relative to its center. Then holds:

$$\sin \vartheta_t = \frac{2\sqrt{2}}{3} \quad \text{and} \quad \cos \vartheta_t = -\frac{1}{3}. \tag{3.70}$$

Proof. Let O' be the projection of the tetrahedron center O on one of the faces with vertices A, B and C (Fig. 3.3(a)). Altitude $O'H$ of the triangle $AO'C$ is simultaneously a bisector, hence holds:

$$\frac{a}{2b} = \sin \frac{\pi}{3}. \tag{3.71}$$

In Figure 3.3(b) we see that

$$\theta_t := \pi - 2\alpha$$

and note that

$$\sin \alpha = \frac{b}{a}. \tag{3.72}$$

Since

$$\sin \frac{\pi}{3} = \frac{\sqrt{3}}{2},$$

from (3.71) and (3.72) follows that

$$\sin\alpha = \frac{1}{\sqrt{3}}, \quad \cos\alpha = \sqrt{1-\sin^2\alpha} = \sqrt{\frac{2}{3}},$$

therefore

$$\sin\theta_t = \sin(\pi - 2\alpha) = \sin(2\alpha) = 2\sin\alpha\cos\alpha = \frac{2\sqrt{2}}{3}$$

and

$$\cos\theta_t = -\sqrt{1-\sin^2\theta_t} = -\frac{1}{3}. \quad \square$$

We know that peptide groups are mostly fixed in the *trans* conformation, i.e., the related primary dihedral angles are equal to π (see Subsection 1.6.1). Thus, we consider the distance l_α between α-carbons of two consequently connected residues to be fixed and equal for all residue pairs. The following theorem permits an evaluation of the distance between two remote α-carbons.

THEOREM 3.20

Let $A_{C_\alpha^i}$, $A_{C_\alpha^{i+1}}$, $\ldots$, $A_{C_\alpha^{i+n}}$ be centers of α-carbons in consequently connected residues numbered i, $i+1$, $\ldots$, $i+n$, and let N^j, $(C')^j$ denote the numbers of the nitrogen and carboxyl carbon of the main chain in the j-th residue ($j = \overline{i,i+n}$) respectively. Then the distance between $A_{C_\alpha^i}$ and $A_{C_\alpha^{i+n}}$ is given by:

$$\left\|\vec{\mathbf{r}}_{C_\alpha^i C_\alpha^{i+n}}\right\| = \sqrt{\left\|\vec{\mathbf{r}}_{C_\alpha^i C_\alpha^{i+n-1}}\right\|^2 + l_\alpha^2 + 2\,\vec{\mathbf{c}}^{\mathrm{T}} \left(\sum_{j=i+1}^{i+n-1}\prod_{m=j}^{i+n-1}\mathbf{R}_m^{[C_\alpha]}\right)\vec{\mathbf{c}}}, \tag{3.73}$$

where

$$\vec{\mathbf{c}} := \begin{pmatrix} l_{C_\alpha^i (C')^i} + \frac{1}{3}l_{(C')^i N^{i+1}} + l_{N^{i+1} C_\alpha^{i+1}} \\ 0 \\ -\frac{\sqrt{3}}{2}l_{(C')^i N^{i+1}} \end{pmatrix}^{\mathrm{T}}, \tag{3.74}$$

and

$$\mathbf{R}_m^{[C_\alpha]} := \begin{pmatrix} \frac{1}{3} & \frac{2\sqrt{2}}{3}\sin\psi & \frac{2\sqrt{2}}{3}\cos\psi \\ -\frac{2\sqrt{2}}{3}\sin\phi & -\cos\phi\cos\psi + \frac{1}{3}\sin\phi\sin\psi & \cos\phi\sin\psi + \frac{1}{3}\sin\phi\cos\psi \\ \frac{2\sqrt{2}}{3}\cos\phi & -\sin\phi\cos\psi - \frac{1}{3}\cos\phi\sin\psi & \sin\phi\sin\psi - \frac{1}{3}\cos\phi\cos\psi \end{pmatrix} \tag{3.75}$$

with

$$\phi = \phi_m, \quad \psi = \psi_m, \quad m = \overline{i+1, i+n-1}.$$

Proof. Similarly to (3.66) we have:

$$\left\|\vec{\mathbf{r}}_{C_\alpha^i C_\alpha^{i+n}}\right\| = \sqrt{\left\|\vec{\mathbf{r}}_{C_\alpha^i C_\alpha^{i+n-1}}\right\|^2 + l_\alpha^2 + 2\sum_{j=i}^{i+n-2}\vec{\mathbf{r}}_{C_\alpha^j C_\alpha^{j+1}}\cdot\vec{\mathbf{r}}_{C_\alpha^{i+n-1}C_\alpha^{i+n}}}, \tag{3.76}$$

Let $\mathcal{B}_{C_\alpha^j}$ denote the basis $\vec{e}_{C_\alpha^j(C')^j}$, $\vec{e}_{C_\alpha^j(C')^j N^{j+1}}$, $\vec{e}^{\perp}_{C_\alpha^j(C')^j(N^{j+1})}$, $\forall j = \overline{i, i+n-1}$.

Note that

$$\vec{r}_{C_\alpha^j C_\alpha^{j+1}} = \vec{r}_{C_\alpha^j(C')^j} + \vec{r}_{(C')^j N^{j+1}} + \vec{r}_{N^{j+1} C_\alpha^{j+1}}, \quad \forall j = \overline{i, i+n-1}.$$

Substitution of values $\frac{2\pi}{3}$ for $\alpha_{(C')^j}$ and $\alpha_{N^{j+1}}$, as well as π for the dihedral angle into (3.6) shows that $\vec{r}_{C_\alpha^j(C')^j} \upuparrows \vec{r}_{N^{j+1}C_\alpha^{j+1}}$. Utilizing additionally equation (3.27), we obtain representation of $\vec{r}_{C_\alpha^j C_\alpha^{j+1}}$ in the basis $\mathcal{B}_{C_\alpha^j}$ (cf. (3.4)):

$$\vec{r}_{C_\alpha^j C_\alpha^{j+1}} = l_{C_\alpha^j(C')^j}\vec{e}_{C_\alpha^j(C')^j} + l_{(C')^j N^{j+1}}\left(\frac{1}{3}\vec{e}_{C_\alpha^j(C')^j} - \frac{\sqrt{3}}{2}\vec{e}^{\perp}_{C_\alpha^j(C')^j(N^{j+1})}\right) +$$

$$+ l_{N^{j+1}C_\alpha^{j+1}}\vec{e}_{C_\alpha^j(C')^j} =$$

$$= \left(l_{C_\alpha^j(C')^j} + \frac{1}{3}l_{(C')^j N^{j+1}} + l_{N^{j+1}C_\alpha^{j+1}}\right)\vec{e}_{C_\alpha^j(C')^j} - \frac{\sqrt{3}}{2}l_{(C')^j N^{j+1}}\vec{e}^{\perp}_{C_\alpha^j(C')^j(N^{j+1})}. \quad (3.77)$$

Therefore,

$$\Xi_{\mathcal{B}_{C_\alpha^j}}(\vec{r}_{C_\alpha^j C_\alpha^{j+1}}) = \begin{pmatrix} l_{C_\alpha^j(C')^j} + \frac{1}{3}l_{(C')^j N^{j+1}} + l_{N^{j+1}C_\alpha^{j+1}} \\ 0 \\ -\frac{\sqrt{3}}{2}l_{(C')^j N^{j+1}} \end{pmatrix} = \vec{c}^*, \; \forall j = \overline{i, i+n-1}.$$

$$(3.78)$$

Hence, according to Remark 3.15 and equation 3.76 holds:

$$\left\|\vec{r}_{C_\alpha^i C_\alpha^{i+n}}\right\| = \sqrt{\left\|\vec{r}_{C_\alpha^i C_\alpha^{i+n-1}}\right\|^2 + l_\alpha^2 + 2\sum_{j=i}^{i+n-2}\vec{c}^{\mathrm{T}}\,\Xi_{\mathcal{B}_{C_\alpha^j}}\circ\Xi^{-1}_{\mathcal{B}_{C_\alpha^{i+n-1}}}(\vec{c})}. \quad (3.79)$$

By induction we deduce:

$$\vec{c}^{\mathrm{T}}\,\Xi_{\mathcal{B}_{C_\alpha^j}}\circ\Xi^{-1}_{\mathcal{B}_{C_\alpha^{i+n-1}}}(\vec{c}) =$$

$$= \vec{c}^{\mathrm{T}}\,\mathbf{A}^{C_\alpha^j C_\alpha^{j+1}}_{C'(trans)^j N^{j+1}}\mathbf{A}^{C'(trans)^j C'(trans)^{j+1}}_{N^{j+1} C_\alpha^{j+1}}\mathbf{A}^{N^{j+1} N^{j+2}}_{C_\alpha^{j+1} C'(trans)^{j+1}}\cdots$$

$$\cdots\mathbf{A}^{C_\alpha^{i+n-2} C_\alpha^{i+n-1}}_{C'(trans)^{i+n-2} N^{i+n-1}}\mathbf{A}^{C'(trans)^{i+n-2} C'(trans)^{i+n-1}}_{N^{i+n-1} C_\alpha^{i+n-1}}\mathbf{A}^{N^{i+n-1} N^{i+n}}_{C_\alpha^{i+n-1} C'(trans)^{i+n-1}}\vec{c} =$$

$$= \vec{c}^{\mathrm{T}}\,\mathbf{R}^{[C_\alpha]}_{j+1}\cdots\mathbf{R}^{[C_\alpha]}_{i+n-1}\vec{c},$$

where

$$\mathbf{R}^{[C_\alpha]}_m := \mathbf{A}^{C_\alpha^{m-1} C_\alpha^m}_{C'(trans)^{m-1} N^m}\mathbf{A}^{C'(trans)^{m-1} C'(trans)^m}_{N^m C_\alpha^m}\mathbf{A}^{N^m N^{m+1}}_{C_\alpha^m C'(trans)^m},$$

$$\forall m = \overline{j+1, i+n-1}.$$

---

*The bond lengthes $l_{C_\alpha^j(C')^j}$, $l_{(C')^j N^{j+1}}$, and $l_{N^{j+1}C_\alpha^{j+1}}$ are the same for all $j = \overline{i, i+n-1}$.

Substituting the above discussed values for fixed dihedral and bond angles, and utilizing conventional notations for variable dihedral angles of the main chain, we obtain:

$$\mathbf{A}^{C'(trans)^{m-1}\,C'(trans)^m}_{N^m\,C^m_\alpha} = \begin{pmatrix} \frac{1}{2} & \frac{\sqrt{3}}{2}\sin\phi_m & \frac{\sqrt{3}}{2}\cos\phi_m \\ 0 & \cos\phi_m & -\sin\phi_m \\ -\frac{\sqrt{3}}{2} & \frac{1}{2}\sin\phi_m & \frac{1}{2}\cos\phi_m \end{pmatrix}, \tag{3.80}$$

$$\mathbf{A}^{N^m\,N^{m+1}}_{C^m_\alpha\,C'(trans)^m} = \begin{pmatrix} \frac{1}{3} & \frac{2\sqrt{2}}{3}\sin\psi_m & \frac{2\sqrt{2}}{3}\cos\psi_m \\ 0 & \cos\psi_m & -\sin\psi_m \\ -\frac{2\sqrt{2}}{3} & \frac{1}{3}\sin\psi_m & \frac{1}{3}\cos\psi_m \end{pmatrix}, \tag{3.81}$$

$$\mathbf{A}^{C^m_\alpha\,C^{m+1}_\alpha}_{C'(trans)^m\,N^{m+1}} = \begin{pmatrix} \frac{1}{2} & 0 & -\frac{\sqrt{3}}{2} \\ 0 & -1 & 0 \\ -\frac{\sqrt{3}}{2} & 0 & -\frac{1}{2} \end{pmatrix}, \tag{3.82}$$

$$\mathbf{A}^{C^m_\alpha\,C^{m+1}_\alpha}_{C'(cis)^m\,N^{m+1}} = \begin{pmatrix} \frac{1}{2} & 0 & \frac{\sqrt{3}}{2} \\ 0 & 1 & 0 \\ -\frac{\sqrt{3}}{2} & 0 & \frac{1}{2} \end{pmatrix}. \tag{3.83}$$

From (3.79)-(3.83) follows (3.73) with (3.75). $\square$

3.3 Energy as a Function of Torsion Angles

Let the molecule be represented by a bidirectional weighted connected graph (see Figure 3.4). The vertices of the graph correspond to atoms, and the edges — to bonds. A covalent bond, e.g., between atoms A_i and A_j, is reproduced two times: as bond b_{ij} of atom A_i to atom A_j, and as bond b_{ji} of atom A_j to A_i. Let us consider bond b_{ij} to be *directed towards* A_j and call b_{ji} *complementary* to b_{ij}.

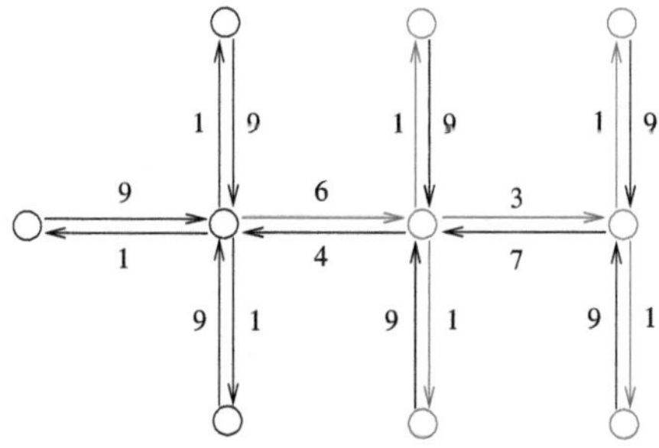

FIG. 3.4: A graph of a molecule with assigned bond weights. Red arrows stand for the bonds in the direction of flow from the bond with the weight six. Red circles denote the atoms that belong to the branch, originated by that bond.

DEFINITION 3.21

Assume there exist a sequence $S = (i_1, \ldots, i_n)$ of atom numbers, such that $i_1 \bowtie i_n$ and $i_j \bowtie i_{j+1}$ for $j = \overline{1, n-1}$. Let the corresponding set

$$\mathcal{R} := \{A_k \mid k \in S\} \cup \{b_{ij} \mid (i, j \in S) \wedge (i \bowtie j)\}$$

be termed as a *ring*, and let $\widehat{\mathcal{R}}_{ij}$ denote the union of all rings containing b_{ij}. A *single* ring includes exactly two bonds of any atom that belongs to it. A *double* ring is a union of two single rings that contain exactly one pair of mutually complementary bonds in common. Other types of rings are not represented in amino acids.

DEFINITION 3.22

Let $\mathcal{F}_{ij}$ denote a set, such that

    a) $b_{ij} \in \mathcal{F}_{ij}$,

    b) if $b_{kl} \in \mathcal{F}_{ij}$, then $b_{lm} \in \mathcal{F}_{ij}$, $\forall k, l, m \in \mathcal{N}$, such that $(k \bowtie l) \wedge (l \bowtie m)$,
                unless $(b_{ml} \in \mathcal{F}_{ij}) \wedge (\widehat{\mathcal{R}}_{lm} = \emptyset)$,

and let it be referred as *bonds in the direction of flow from* b_{ij} (see Figure 3.4).

DEFINITION 3.23

Let $\mathcal{D}_{ij} := \{A_k \mid b_{lk} \in \mathcal{F}_{ij}, \ \forall k, l \in \mathcal{N}\}$ be termed as the *branch, originated by* b_{ij}, A_i be referred as the *origin* of branch $\mathcal{D}_{ij}$, and let branches $\mathcal{D}_{ij}$, $\mathcal{D}_{ji}$ be called *complementary*. Besides, let us denote $\mathcal{N}_{jk} := \{n \in \mathcal{N} \mid A_n \in \mathcal{D}_{jk}\}$.

A twisting of a chain at the covalent bond between atoms A_i and A_j can be achieved by means of rotation of either branch $\mathcal{D}_{ij}$ or branch $\mathcal{D}_{ji}$. However, it is more efficient to rotate the branch that contains less atoms. Therefore, a *bond weight* w_{ij} is assigned to each bond b_{ij}:

$$w_{ij} = \begin{cases} 0, & \text{if } \widehat{\mathcal{R}}_{ij} \neq \emptyset, \\ \operatorname{card} \mathcal{D}_{ij} - 1 & \text{otherwise.} \end{cases} \tag{3.84}$$

Now, let

$$\mathcal{Z} := \{(i, j) \in \mathcal{N}^2 \mid (i \bowtie j) \wedge (w_{ij} \neq 1) \wedge (w_{ji} \neq 1)\},$$
$$\mathcal{Y} := \{(i, j) \in \mathcal{Z} \mid (i \simeq j) \vee (\widehat{\mathcal{R}}_{ij} \neq \emptyset)\},$$
$$\mathcal{X} := \mathcal{Z} \setminus \mathcal{Y}, \text{ and } M := \frac{1}{2} \operatorname{card} \mathcal{X}. \tag{3.85}$$

Further, let

$$\mathcal{M} := \{1, 2, \ldots, M\},$$

and let $\kappa : \mathcal{X} \to \mathcal{M}$ be a surjection, such that

$$\kappa(i, j) = \kappa(j, i), \text{ and } \kappa(i, j) \neq \kappa(k, l) \text{ unless } (i, j) = (k, l) \text{ or } (i, j) = (l, k).$$

Thus, κ can be understood as a numbering for degrees of freedom.

Besides, let

$$\mathcal{C} := \{(i,j) \in \mathcal{N}^2 \mid (i \nsim j) \wedge (i < j)\}, \tag{3.86}$$

and let $\vec{\mathbf{I}} : \mathcal{C} \to \mathcal{N}^N$ be a map, such that

$$I^{[1]}(i,j) = i,$$
$$I^{[k]} \bowtie I^{[k+1]}, \ \forall k = \overline{1, \aleph_{ij} + 1},$$
$$I^{[k]} = j, \ \forall k = \overline{\aleph_{ij} + 2, N}.$$

Hence, for any ordered couple of atom indexes, such that $i \nsim j$, I is a vector containing the sequence of atom numbers that corresponds to the shortest connected path from A_i to A_j. The path is not always uniquely defined due to the presence of rings.

To model the dependence of the free energy on the variable torsion angles in a molecule, we define a map $\widetilde{U} : (-\pi, \pi]^M \to \mathbb{R}$ as:

$$\widetilde{U}(\vec{\zeta}) = \widetilde{U}^{[w]}(\vec{\zeta}) + \widetilde{G}^{[e]}(\vec{\zeta}) + \widetilde{G}^{[s]}(\vec{\zeta}), \tag{3.87}$$

where

$$\widetilde{U}^{[w]}(\vec{\zeta}) := \sum_{\substack{(i,j)\in\mathcal{C} \\ d_{ij}(\vec{\zeta})<C^{[w]}}} \sqrt{E_i^{[w]} E_j^{[w]}} \left(\left(\frac{R_i^{[w]} + R_j^{[w]}}{d_{ij}(\vec{\zeta})} \right)^{12} - 2 \left(\frac{R_i^{[w]} + R_j^{[w]}}{d_{ij}(\vec{\zeta})} \right)^6 \right) \tag{3.88}$$

is the potential energy of intramolecular van der Waals interactions (cf. (2.7)), which is to be modified in case of hydrogen bonding, $C^{[w]}$ is the cutoff used for computations of these interactions,

$$\widetilde{G}^{[e]}(\vec{\zeta}) := \frac{1}{4\pi\epsilon_0} \sum_{(i,j)\in\mathcal{C}} \frac{q_i q_j}{\epsilon_{ij}(\vec{\zeta}) d_{ij}(\vec{\zeta})} \tag{3.89}$$

is the free energy of electrostatic interactions between protein atoms, $\epsilon_{ij}(\vec{\zeta})$ and $\widetilde{G}^{[s]}(\vec{\zeta})$ are respectively the screening functions and the free energy of solvation, both described in Section 3.4, and $d_{ij} : (-\pi, \pi]^M \to \mathbb{R}^+$ are distances between atoms A_i and A_j ($(i,j) \in \mathcal{C}$):

$$d_{ij}(\vec{\zeta}) := \left(\sum_{k=2}^{\aleph_{ij}+2} l^2_{I^{[k-1]}(i,j)\, I^{[k]}(i,j)} + \right.$$

$$\left. +2 \sum_{k=2}^{\aleph_{ij}+1} \sum_{n=k}^{\aleph_{ij}+1} l_{I^{[k-1]}(i,j)\, I^{[k]}(i,j)}\, l_{I^{[n]}(i,j)\, I^{[n+1]}(i,j)} \left(\prod_{m=k}^{n} \mathbf{A}^{I^{[m-1]}(i,j) I^{[m+2]}(i,j)}_{I^{[m]}(i,j) I^{[m+1]}(i,j)} \right)_{11} \right)^{1/2}. \tag{3.90}$$

Here $\mathbf{A}^{I^{[m-1]}(i,j) I^{[m+2]}(i,j)}_{I^{[m]}(i,j) I^{[m+1]}(i,j)}$ given by (3.64) for $m = \overline{2, \aleph_{ij}}$, and by (3.65) for $m = \aleph_{ij}+1$.

$\mathbf{A}_{I^{[m]}(i,j)I^{[m+1]}(i,j)}^{I^{[m-1]}(i,j)I^{[m+2]}(i,j)}$ is a constant matrix, if $m = \aleph_{ij}+1$ or $(I^{[m]}(i,j), I^{[m+1]}(i,j)) \in \mathcal{Y}$, and a function of $\zeta^{[\kappa(I^{[m]}(i,j),I^{[m+1]}(i,j))]}$ otherwise. In the latter case

$$\zeta_{I^{[m]}(i,j)I^{[m+1]}(i,j)}^{I^{[m-1]}(i,j)I^{[m+2]}(i,j)} = \zeta^{[\kappa(I^{[m]}(i,j),I^{[m+1]}(i,j))]}+$$

$$+ \zeta_s(I^{[m-1]}(i,j), I^{[m]}(i,j), I^{[m+1]}(i,j), I^{[m+2]}(i,j)),$$

where $\zeta_s(I^{[m-1]}(i,j), I^{[m]}(i,j), I^{[m+1]}(i,j), I^{[m+2]}(i,j)) \in (-\pi, \pi)$ is a constant supplement that depends on the choice of the primary chain and the chiral configurations of the centers $A_{I^{[m]}(i,j)}$ and $A_{I^{[m+1]}(i,j)}$.

REMARK 3.24
Note that $d_{ij}(\vec{\zeta}) = d_{ji}(\vec{\zeta})$, $\forall i, j \in \mathcal{N}$. The condition of number ordering in (3.86) is introduced to avoid double inclusion of energies related to each atom pair into the estimated potential of mean force $\widetilde{U}(\vec{\zeta})^*$.

3.4   MODELING HYDRATION

The implicit solvation models, described in Chapter 2, are based of the observation of the linear correlation between hydrophobic effect and SASA. However, since the number of water molecules that can directly contact an atom is discrete, it can be of advantage to estimate this number instead of more exact computation or estimation of the solvent-exposed surface area. Thus, it is not unreasonable to assume that the effect of the interaction of one water molecule with a cavity on solute surface remains the same after doubling the cavity SASA, if the size of the cavity is still not sufficient to accommodate more molecules. The idea to estimate the number of contacted water molecules for each atom is set as a basis for the below described method.

For this purpose, a grid consisting of twelve uniformly distributed points is generated on the surface of each atom (Fig. 3.5(a)). These points describe potential locations of water molecules closely packed around an isolated atom. The point positions are updated periodically through a random rotation of the grid around the atom center[†]. To estimate the *hydration degree* h_i of atom A_i, each grid point is checked with regard to its accessibility by water, and the corresponding hydration status is assigned to it (Fig. 3.5(b)). The number of accessible points $n_i^{[a]}$ is used for evaluation of the solvation energy $G_i^{[s]}$ of atom A_i:

$$G_i^{[s]} = E_i^{[s]} n_i^{[a]}. \tag{3.91}$$

Here $E_i^{[s]}$ is an estimated solvation energy per one contacted water molecule, depending on the atom type and charge. The hydration degree h_i is then $n_i^{[a]}/12$.

---

[*]$\widetilde{U}(\vec{\zeta})$ has free energy components, but by itself is not a free energy, since it does not take into account the entropy of the chain.

[†]In fact, for more computational efficiency the surrounding coordinate space is rotated instead of the surface grid, as described below.

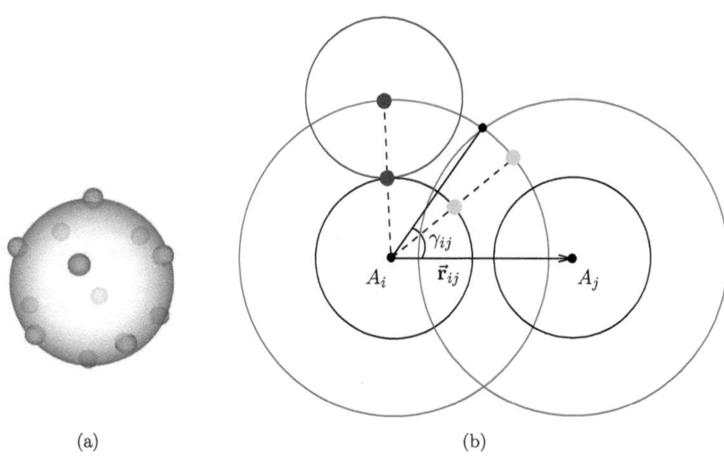

FIG. 3.5: (a) An isolated atom with its solvation grid. (b) Two atoms with overlapping hydration spheres (red). Grid points that lay in the intersection cone are marked as inaccessible (light orange). Initially all points are considered to be accessible (blue). The points on hydration spheres describe potential locations of water molecules, and the points on the van der Waals surface are utilized for visualization of surface accessibility (see also Figure 3.10). The both related points correspond to the same grid point position, described by its surface coordinates.

To account for the screening effect, the relative permittivity $\epsilon_{ij} = 40$ is used for evaluation of electrostatic interactions, if at least one of the two interacting atoms A_i and A_j is hydrated. Otherwise, $\epsilon_{ij} = 3$ is taken, largely in accordance with justifications given in [1]*.

3.4.1 RATIONALE

One can place around a sphere S of the radius R at maximum 12 spheres of the same radius. This problem has a long history, being already a subject of discussion between Isaac Newton and David Gregory in Cambridge [117]. However, the first proof that is accepted nowadays was given only in 1953 by Schütte and van der Waerden [118]. Since there is no need for a rigorous estimate in case of our approximation, we shall restrict ourselves with the examination described below.

Let us explore a regular icosahedron (Fig. 3.6(a)), inscribed in a sphere with the radius $2R$ around the center of the sphere S.

---

*Hereby we make simplifying assumptions that any non-hydrated charges are closer to each other then to the protein surface and that the charges at the opposite sides of the protein surface are separated by a sufficient distance to be essentially invisible to each other even at $\epsilon_{ij} = 40$.

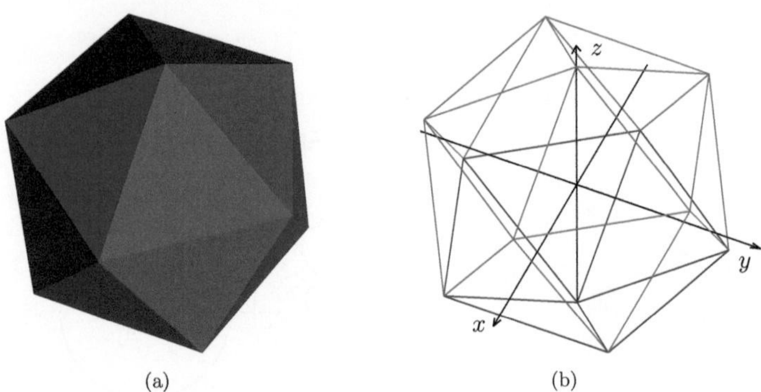

(a)   (b)

Fig. 3.6: (a) An opaque icosahedron. (b) Orientation of an icosahedron in the coordinate system used for generation of initial surface grid coordinates. The faces situated over the xy-plane are outlined by red color, and those under xy-plane — by blue.

Lemma 3.25
Let a be the length of the ridge of a regular icosahedron, and $\hat{R}$ be the radius of the circumscribed sphere. Then holds:

$$a = 2\hat{R}\sqrt{\frac{5 - \sqrt{5}}{10}} \approx 1.05\hat{R}. \tag{3.92}$$

Proof. Consider a section through five icosahedron vertices (see Figure 3.7 (a) and notations therein). Note that

$$\alpha = \frac{2\pi}{5}, \quad \beta = \frac{1}{2}(\pi - \alpha) = \frac{3\pi}{10}, \quad \text{and} \quad \gamma = \frac{\pi}{2} - \beta = \frac{\pi}{5}.$$

By Ptolemy's theorem [119], which says that the product of diagonals of a cyclic quadrilateral is equal to the sum of the products of the opposite sides, it follows that a and b are in golden ratio:

$$AD \cdot CE = AE \cdot CD + DE \cdot AC \Rightarrow b^2 = a^2 + ab \Rightarrow \frac{b}{a} = \frac{1 + \sqrt{5}}{2}.$$

From the other side, we can see in triangle BCF that

$$\cos\gamma = \frac{b}{2a},$$

hence

$$\cos\frac{\pi}{5} = \frac{1 + \sqrt{5}}{4} \quad \text{and} \quad \sin\frac{\pi}{5} = \sqrt{1 - \cos^2\frac{\pi}{5}} = \frac{1}{2}\sqrt{\frac{5 - \sqrt{5}}{2}}. \tag{3.93}$$

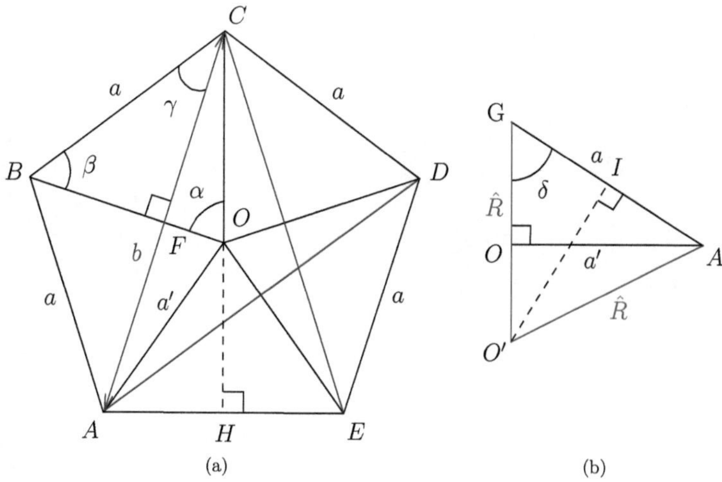

FIG. 3.7: (a) A cross-section through five icosahedron vertices laying in one plane. $AB = BC = CD = DE = AE =: a$, $AO = BO = CO = DO = EO =: a'$, $AC = CE = AD =: b$ (i.a.), $\angle AOB = \angle BOC = \angle COD = \angle DOE = \angle AOE =: \alpha$, $\angle OBC = \angle BCO =: \beta$ (i.a.), $\angle BCF =: \gamma$ (i.a.). (b) A cross-section through two vertices and the center O' of the icosahedron. $AG = a$, $AO = a'$, $O'A = O'G = \hat{R}$, $\angle AGO' = \angle GAO' =: \delta$.

Altitude OH of triangle AOE is simultaneously a bisector, therefore holds:

$$\frac{a}{2a'} = \sin\frac{\pi}{5},$$

and using (3.93) we obtain:

$$\frac{a'}{a} = \sqrt{\frac{5 + \sqrt{5}}{10}}. \tag{3.94}$$

Now consider a cross-section through two vertices and the center of the icosahedron (Fig. 3.7 (b)). We see from OGA that

$$\sin\delta = \frac{a'}{a}, \tag{3.95}$$

and from $O'GI$

$$\cos\delta = \frac{a}{2\hat{R}}. \tag{3.96}$$

Note that according to (3.94) and (3.95)

$$\cos\delta = \sqrt{1 - \left(\frac{a'}{a}\right)^2} = \sqrt{\frac{5 - \sqrt{5}}{10}}, \tag{3.97}$$

and with (3.96) immediately follows (3.92). $\square$

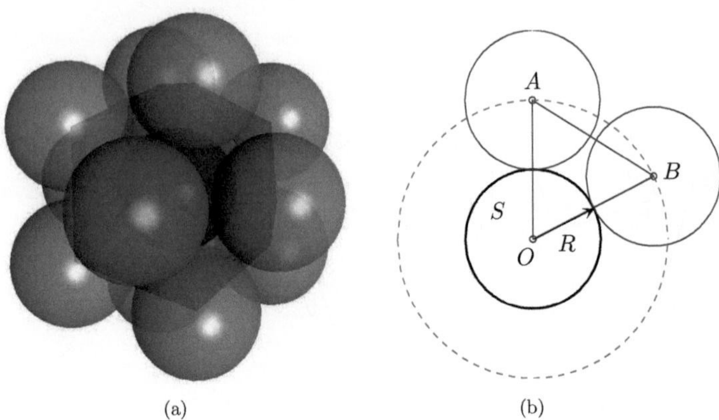

(a)   (b)

Fig. 3.8: (a) Placement of twelve blue spheres in icosahedron vertices about one black sphere in the center. (b) A cross section through the center of the sphere S of radius R and two vertices A and B of the discussed icosahedron.

We see that the length of the ridge of the given icosahedron is approximately 1.05 times larger than the distance from a vertex to the icosahedron center, equal to $2R$, which means that spheres of radius R placed at the vertices connected by a ridge almost touch each other (Fig. 3.8(a,b)).

The conventional probe radius R_p roughly corresponds to an average van der Waals radius of atoms constituting typical organic molecules, in particular, proteins. Therefore, taking into account an approximate nature of the definition of solvent accessible surface, we assume that about 12 water molecules can be placed around an isolated atom, neglecting for simplicity the difference of van der Waals radii.

Let $R_i^{[w]}$ be the van der Waals radius of atom A_i, and let us denote the sphere of radius $R_i^{[h]} := R_i^{[w]} + R_p$ around the atom center as a *hydration sphere* of the atom. To determine, which of 12 potential positions of water molecules are in fact accessible, a grid of 12 uniformly distributed points is generated on the surface of the atom hydration sphere. The location of each grid point is described by its surface coordinates (φ, ϑ) defined as in Figure 3.9.

3.4.2   Grid generation

Using the fact that the grid points correspond to the vertices of the inscribed icosahedron (Fig. 3.5(a)), we initialize the grid on the stage of creation by the following coordinates (see Figure 3.6(b)):

$$(0,0),\ (0,\vartheta_I),\ \left(\tfrac{2\pi}{5},\vartheta_I\right),\ \left(-\tfrac{2\pi}{5},\vartheta_I\right),\ \left(\tfrac{4\pi}{5},\vartheta_I\right),\ \left(-\tfrac{4\pi}{5},\vartheta_I\right), \tag{3.98}$$

$$\left(\tfrac{\pi}{5},\pi-\vartheta_I\right),\ \left(-\tfrac{\pi}{5},\pi-\vartheta_I\right),\ \left(\tfrac{3\pi}{5},\pi-\vartheta_I\right),\ \left(-\tfrac{3\pi}{5},\pi-\vartheta_I\right),\ (\pi,\pi-\vartheta_I),\ (0,\pi), \tag{3.99}$$

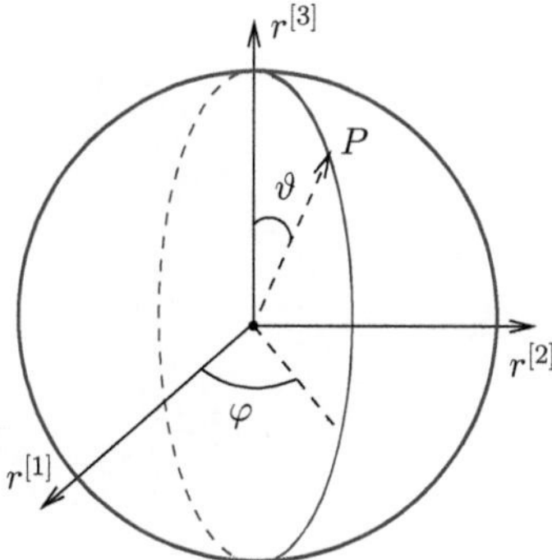

FIG. 3.9: Surface coordinates (φ, ϑ) of location of the point P on the surface of the atom sphere and their relation to the Cartesian coordinates.

where ϑ_I is given by the following lemma.

LEMMA 3.26
For the angular distance ϑ_I between two vertices of a regular icosahedron relative to its center holds:

$$\vartheta_I = \arctan 2. \tag{3.100}$$

Proof. From Figure 3.7(b) we see that

$$\sin \frac{\vartheta_I}{2} = \cos \delta.$$

Therefore

$$\cos \vartheta_I = 1 - 2 \cos^2 \delta,$$

and with (3.97) we obtain

$$\cos \vartheta_I = \frac{1}{\sqrt{5}}.$$

Since for any α holds

$$1 + \tan^2 \alpha = \frac{1}{\cos^2 \alpha},$$

immediately follows (3.100). $\square$

In order to obtain an arbitrary oriented grid, at the beginning of each energy minimization step a new rotation matrix

$$\mathbf{M}_i = \begin{pmatrix} \cos\psi\cos\varphi - \cos\vartheta\sin\psi\sin\varphi & \sin\psi\cos\varphi + \cos\vartheta\cos\psi\sin\varphi & \sin\vartheta\sin\varphi \\ -\cos\psi\sin\varphi - \cos\vartheta\sin\psi\cos\varphi & -\sin\psi\sin\varphi + \cos\vartheta\cos\psi\cos\varphi & \sin\vartheta\cos\varphi \\ \sin\vartheta\sin\psi & -\sin\vartheta\cos\psi & \cos\vartheta \end{pmatrix}$$

is computed for each atom A_i from a triple $(\vartheta, \psi, \varphi) \in [0, \pi) \times [0, \pi) \times [0, 2\pi)$ of randomly generated Euler angles [120] and stored in the atom class. This matrix

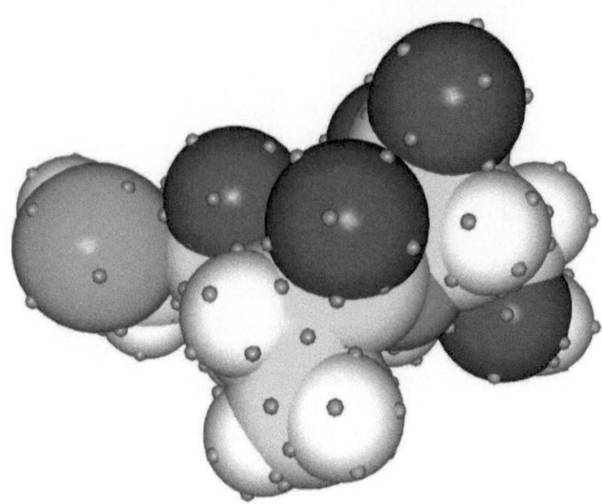

Fig. 3.10: Cysteine-alanine-serine tripeptide with its solvation grid. Hydrated points are colored in light blue, and the grid points laying in cavities are colored in orange.

can be used for rotation of the surface grid, as shown in Figure 3.10. However, it is more efficient to rotate the surrounding coordinate space, i.e. compute the vectors $\mathbf{M}_i\vec{\mathbf{r}}_{ij}$ and $\mathbf{M}_j\vec{\mathbf{r}}_{ji}$ for each pair of atoms A_i and A_j with intersecting hydration spheres, and store only one set of initial grid coordinates, given by (3.98)-(3.99), for all atoms.

3.4.3 Accessibility check

Definition 3.27
Let $\Upsilon : \mathbb{R} \times (-\pi, \pi] \times [0, \pi] \to \mathbb{R}^3$ be a mapping of spherical to Cartesian coordinates, defined as

$$\Upsilon(r, \varphi, \vartheta) := \begin{pmatrix} r \sin \vartheta \cos \varphi \\ r \sin \vartheta \sin \varphi \\ r \cos \vartheta \end{pmatrix}. \tag{3.101}$$

Let the inverse mapping $\Upsilon^{-1}(\vec{\mathbf{r}})$ be defined as specified in (1.23)-(1.25), and let $\Upsilon^{-1}_{[\varphi,\vartheta]} : \mathbb{R}^3 \to (-\pi, \pi] \times [0, \pi]$ denote the projection of Cartesian onto surface coordinates (φ, ϑ), given by (1.24)-(1.25).

Definition 3.28
Let

$$\Gamma_i := \left\{ \vec{\mathbf{p}} \in \mathbb{R}^3 \mid \|\vec{\mathbf{p}} - \vec{\mathbf{r}}_i\| = R_i^{[h]} \right\} \text{ and } \gamma_{ij} := \angle(\vec{\mathbf{g}} - \vec{\mathbf{r}}_i, \vec{\mathbf{r}}_{ij}), \ \vec{\mathbf{g}} \in \Gamma_i \cap \Gamma_j. \tag{3.102}$$

We say that a grid point (φ_k, ϑ_k) is inside the intersection of hydration spheres of atoms A_i and A_j if and only if holds:

$$\left(\vec{\mathbf{r}}_i + \Upsilon(R_i^{[h]}, \varphi_k, \vartheta_k) \right) \in \Lambda_{ij} := \{\vec{\mathbf{p}} \in \Gamma_i \mid \angle(\vec{\mathbf{p}} - \vec{\mathbf{r}}_i, \vec{\mathbf{r}}_{ij}) < \gamma_{ij}\}. \tag{3.103}$$

PROPOSITION 3.29

γ_{ij} *defined in (3.102) is given by*

$$\gamma_{ij} = \arccos\left(\frac{(R_i^{[h]})^2 - (R_j^{[h]})^2 + \|\vec{\mathbf{r}}_{ij}\|^2}{2R_i^{[h]}\|\vec{\mathbf{r}}_{ij}\|}\right). \tag{3.104}$$

Proof. Let $\vec{\mathbf{g}} \in \Gamma_i \cap \Gamma_j$, so that holds:

$$(\vec{\mathbf{g}} - \vec{\mathbf{r}}_l) \cdot (\vec{\mathbf{g}} - \vec{\mathbf{r}}_l) = R_l^{[h]}, \quad l \in \{i, j\}. \tag{3.105}$$

We have:

$$\cos\gamma_{ij} = \cos\angle(\vec{\mathbf{g}} - \vec{\mathbf{r}}_i, \vec{\mathbf{r}}_{ij}) = \cos\left(\frac{(\vec{\mathbf{g}} - \vec{\mathbf{r}}_i) \cdot \vec{\mathbf{r}}_{ij}}{\|\vec{\mathbf{g}} - \vec{\mathbf{r}}_i\|\|\vec{\mathbf{r}}_{ij}\|}\right) = \cos\left(\frac{(\vec{\mathbf{g}} - \vec{\mathbf{r}}_i) \cdot \vec{\mathbf{r}}_{ij}}{R_i^{[h]}\|\vec{\mathbf{r}}_{ij}\|}\right), \tag{3.106}$$

and using (3.105) we obtain:

$$\begin{aligned}
(\vec{\mathbf{g}} - \vec{\mathbf{r}}_i) \cdot \vec{\mathbf{r}}_{ij} &= (\vec{\mathbf{g}} - \vec{\mathbf{r}}_i) \cdot (\vec{\mathbf{r}}_j - \vec{\mathbf{r}}_i) = (\vec{\mathbf{g}} - \vec{\mathbf{r}}_i) \cdot (\vec{\mathbf{r}}_j - \vec{\mathbf{g}} + \vec{\mathbf{g}} - \vec{\mathbf{r}}_i) = \\
&= (\vec{\mathbf{g}} - \vec{\mathbf{r}}_i) \cdot (\vec{\mathbf{r}}_j - \vec{\mathbf{g}}) + (R_i^{[h]})^2 = (\vec{\mathbf{g}} - \vec{\mathbf{r}}_j + \vec{\mathbf{r}}_j - \vec{\mathbf{r}}_i) \cdot (\vec{\mathbf{r}}_j - \vec{\mathbf{g}}) + (R_i^{[h]})^2 = \\
&= -(R_j^{[h]})^2 + \vec{\mathbf{r}}_{ij} \cdot (\vec{\mathbf{r}}_j - \vec{\mathbf{g}}) + (R_i^{[h]})^2 = \\
&= -(R_j^{[h]})^2 + \vec{\mathbf{r}}_{ij} \cdot (\vec{\mathbf{r}}_j - \vec{\mathbf{r}}_i + \vec{\mathbf{r}}_i - \vec{\mathbf{g}}) + (R_i^{[h]})^2 = \\
&= -(R_j^{[h]})^2 + \|\vec{\mathbf{r}}_{ij}\|^2 - \vec{\mathbf{r}}_{ij} \cdot (\vec{\mathbf{g}} - \vec{\mathbf{r}}_i) + (R_i^{[h]})^2.
\end{aligned}$$

Hence

$$(\vec{\mathbf{g}} - \vec{\mathbf{r}}_i) \cdot \vec{\mathbf{r}}_{ij} = \frac{1}{2}\left((R_i^{[h]})^2 - (R_j^{[h]})^2 + \|\vec{\mathbf{r}}_{ij}\|^2\right), \tag{3.107}$$

and a substitution of (3.107) into (3.106) gives (3.104). $\square$

PROPOSITION 3.30

A surface grid point $P_k := (\varphi_k, \vartheta_k)$ of atom A_i is inside the intersection of hydration spheres of atoms A_i and A_j if and only if the following inequality holds:

$$\sin\vartheta_k \sin\vartheta_{ij} \cos(\varphi_k - \varphi_{ij}) + \cos\vartheta_k \cos\vartheta_{ij} > \cos\gamma_{ij}, \tag{3.108}$$

where $(\varphi_{ij}, \vartheta_{ij}) = \Upsilon^{-1}_{[\varphi,\vartheta]}(\vec{\mathbf{r}}_{ij})$.

Proof. The angle between vector $\vec{\mathbf{r}}_{ij}$ and the unit vector $\vec{\mathbf{v}}_k$ pointing from the atom center A_i towards P_k is given by:

$$\begin{aligned}
\angle(\vec{\mathbf{v}}_k, \vec{\mathbf{r}}_{ij}) &= \arccos\left(\frac{\vec{\mathbf{v}}_k \cdot \vec{\mathbf{r}}_{ij}}{\|\vec{\mathbf{r}}_{ij}\|}\right) = \arccos\left(\Upsilon(1, \varphi_k, \vartheta_k) \cdot \Upsilon(1, \varphi_{ij}, \vartheta_{ij})\right) \\
&= \arccos(\sin\vartheta_k \cos\varphi_k \sin\vartheta_{ij} \cos\varphi_{ij} + \\
&\quad + \sin\vartheta_k \sin\varphi_k \sin\vartheta_{ij} \sin\varphi_{ij} + \cos\vartheta_k \cos\vartheta_{ij}) = \\
&= \arccos(\sin\vartheta_k \sin\vartheta_{ij} \cos(\varphi_k - \varphi_{ij}) + \cos\vartheta_k \cos\vartheta_{ij}).
\end{aligned}$$

Since $\angle(\vec{\mathbf{v}}_k, \vec{\mathbf{r}}_{ij})$, $\gamma_{ij} \in [0, \pi]$, the condition $\angle(\vec{\mathbf{v}}_k, \vec{\mathbf{r}}_{ij}) < \gamma_{ij}$ holds if and only if $\cos \angle(\vec{\mathbf{v}}_k, \vec{\mathbf{r}}_{ij}) > \cos \gamma_{ij}$. $\square$

Initially each grid point obtains the status "accessible", i.e. considered to be exposed to solvent. Then for each pair of atoms A_i and A_j, such that

$$\|\vec{\mathbf{r}}_{ij}\| \leq R_i^{[h]} + R_j^{[h]},$$

the surface coordinates

$$(\varphi_{ij}, \vartheta_{ij}) = \Upsilon_{[\varphi, \vartheta]}^{-1} (\mathbf{M}_i \vec{\mathbf{r}}_{ij})$$

of the rotated vector $\vec{\mathbf{r}}_{ij}$ are computed[*], and for each accessible grid point (φ_k, ϑ_k) of atom A_i the test (3.108) is performed. If the point lays inside the intersection, its status is set to "unaccessible". The same is repeated with the roles A_i and A_j exchanged.

3.5 MODELING COTRANSLATIONAL FOLDING

The potential $\widetilde{U}(\vec{\zeta})$ requires a specification of an appropriate initial configuration. As already mentioned in Section 3.1, the strategy consists in generation of optimal coordinates for all constituting amino acids and imitation (with certain simplifications) of protein synthesis as it happens in ribosomes. That is, each new residue is appended in a way that the formed peptide group acquires the *trans* arrangement, and the other dihedral angles obtain suitable[†] initial values.

Since the carboxyl terminal of the growing chain is fixed at PTC and partially remains inside the ribosomal tunnel, at leat some terminal residues can not participate in folding during the synthesis. To imitate this effect, we freeze the last eight[‡] residues and supplement the potential by an additional term:

$$U^{[r]} = \sum_{k=1}^{K-8} \sum_{i \in \mathcal{N}_k} k^{[r]} \min(0, \vec{\mathbf{r}}_{ic} \cdot \vec{\mathbf{v}}_c)^2, \tag{3.109}$$

which is intended to favor folding at one side of the plane orthogonal to the central axis of the frozen helical fragment and passing through $\vec{\mathbf{r}}_c$ (see Figure 3.11). Here K is the current number of residues, $\mathcal{N}_k$ are the numbers of all atoms in the k-th residue, $k^{[r]}$ is a sufficiently large constant, $\vec{\mathbf{r}}_c$ is given by

$$\vec{\mathbf{r}}_c = \frac{1}{4} \sum_{k=K-7}^{K-4} \vec{\mathbf{r}}_{(C_\alpha)_k},$$

---

[*]Here the external coordinate space is rotated instead of the initial surface grid. If A_i has no accessible grid points, this computation is omitted.

[†]For a discussion concerning this issue see Subsections 1.6.1, 1.6.2, and 1.8.3.

[‡]The fragment hidden in the ribosomal tunnel is apparently longer (see Section 1.8), but this number is convenient for determination of an imaginary tunnel axis, both in case of a helical and an extended initial conformation of appended residues. At the same time, this length is sufficient to reproduce the desired effect.

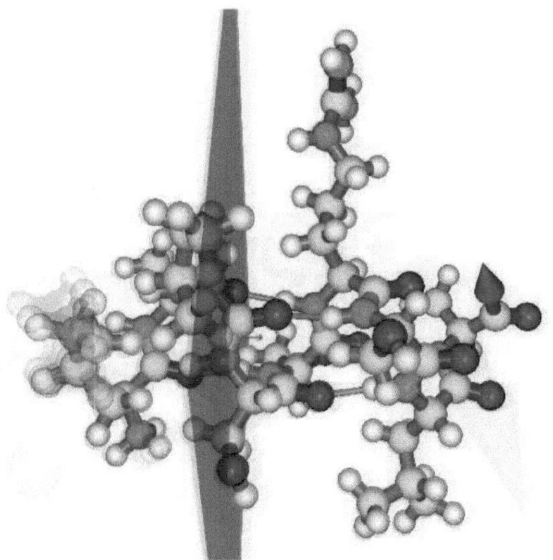

FIG. 3.11: Simulation of cotranslational folding with SIVIPROF. The half-space available for folding is spread at the left-hand side of the blue semitransparent plane. The small pink arrow depicts the normalized $\vec{\mathbf{v}}_c$ pointing from $\vec{\mathbf{r}}_c$, while the orange three-dimensional arrow at the carboxyl end shows the direction of the peptide chain elongation.

and the direction of the putative tunnel axis is given by

$$\vec{\mathbf{v}}_c = \frac{1}{4} \sum_{k=K-3}^{K} \vec{\mathbf{r}}_{(C_\alpha)_k} - \vec{\mathbf{r}}_c.$$

As usually, we imply that $\vec{\mathbf{r}}_{ic} := \vec{\mathbf{r}}_c - \vec{\mathbf{r}}_i$ and $\vec{\mathbf{r}}_{(C_\alpha)_k}$ is the position of the α-carbon belonging to the k-th residue.

REMARK 3.31
It is more efficient to compute $\vec{\mathbf{r}}_s := \vec{\mathbf{r}}_c \cdot \vec{\mathbf{v}}_c$ one time for a given configuration, and then obtain $\vec{\mathbf{r}}_{ic} \cdot \vec{\mathbf{v}}_c$ as $\vec{\mathbf{r}}_s - \vec{\mathbf{r}}_i \cdot \vec{\mathbf{v}}_c$.

REMARK 3.32
$U^{[r]}$ can be also written as a function of dihedral angles $\vec{\zeta}$. Indeed, all atomic positions can be expressed relative to any atom instead of the coordinate origin, and the conformation of the molecule is uniquely defined by the primary dihedral angles.

For generation of initial atomic coordinates and for their subsequent transformations we shall use the operations described in the following subsections. All related algorithms are discussed in Chapter 4.

3.5.1   OPERATIONS FOR COORDINATE TRANSFORMATIONS

In the following derivations we assume that all vectors point from the coordinate origin.

DEFINITION 3.33

Let $\vec{a} \downarrow \alpha$ denote an operation that produces a vector $\vec{a}'$, such that $\angle(\vec{a}', \vec{a}) = \alpha$ and $\|\vec{a}'\| = \|\vec{a}\|$. This operation shall be referred as *deflection* of vector $\vec{a}$ to the angle α. Herewith, however, it is implied that vector $\vec{a}$ itself is not changed, but a new vector is created as the result of this operation.

DEFINITION 3.34

Let $\vec{a} \downarrow^{\vec{b}} \alpha$ denote the result of deflection of vector $\vec{a}$ to the angle α in the plane spanned by non-collinear vectors $\vec{a}$ and $\vec{b}$, such that the vector is rotated clockwise when viewed along the vector $\vec{a} \times \vec{b}$ (Fig. 3.12(a)). Let such operation be also referred as *deflection* to the angle α *in the direction* of vector $\vec{b}$.

DEFINITION 3.35

Let $\vec{a} \,{}^{\alpha}_{\vee}\, \vec{b}$ symbolize the result of deflection of vector $\vec{a}$ in the direction of vector $\vec{b}$, such that the angle between the new vector and vector $\vec{b}$ is equal to α (Fig. 3.12 (b)).

LEMMA 3.36

Let $\vec{a}$ and $\vec{b}$ be non-collinear vectors. The vector $\vec{a}_\perp$ given by

$$\vec{a}_\perp := \vec{b} - \frac{\vec{a}\,(\vec{a} \cdot \vec{b})}{\vec{a} \cdot \vec{a}} \tag{3.110}$$

is orthogonal to $\vec{a}$.

Proof.

$$\vec{a} \cdot \vec{a}_\perp = \vec{a} \cdot \vec{b} - \frac{(\vec{a} \cdot \vec{a})(\vec{a} \cdot \vec{b})}{\vec{a} \cdot \vec{a}} = 0. \;\square$$

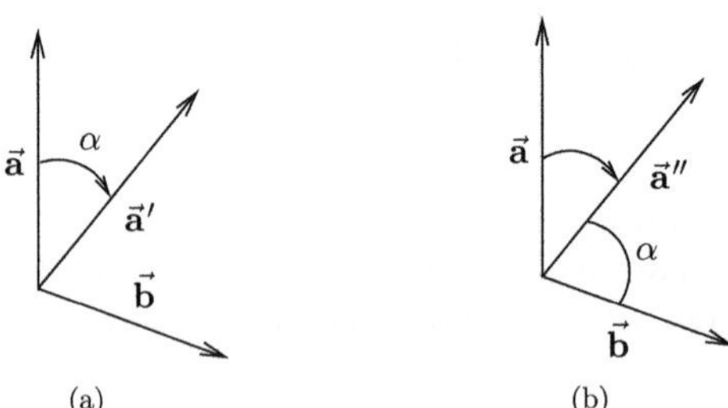

FIG. 3.12: Deflection of the vector $\vec{a}$ in the direction of the vector $\vec{b}$. (a) $\vec{a}' = \vec{a} \downarrow^{\vec{b}} \alpha$. (b) $\vec{a}'' = \vec{a} \,{}^{\alpha}_{\vee}\, \vec{b}$.

PROPOSITION 3.37

Under the conditions of Lemma 3.36 the deflected vector $\vec{\mathbf{a}}' := \vec{\mathbf{a}} \downarrow^{\vec{b}} \alpha$ can be computed using the following expression:

$$\vec{\mathbf{a}}' = \vec{\mathbf{a}} \cos\alpha + \frac{\vec{\mathbf{a}}_\perp \|\vec{\mathbf{a}}\| \sin\alpha}{\|\vec{\mathbf{a}}_\perp\|}.$$

Proof. First, we show that $\vec{\mathbf{a}}'$ is a result of rotation of vector $\vec{\mathbf{a}}$ to the angle α:

$$\|\vec{\mathbf{a}}'\| = \sqrt{\vec{\mathbf{a}}' \cdot \vec{\mathbf{a}}'} = \sqrt{\vec{\mathbf{a}} \cdot \vec{\mathbf{a}} \cos^2\alpha + \frac{2\,\vec{\mathbf{a}} \cdot \vec{\mathbf{a}}_\perp \|\vec{\mathbf{a}}\| \sin\alpha \cos\alpha}{\|\vec{\mathbf{a}}_\perp\|} + \frac{\vec{\mathbf{a}}_\perp \cdot \vec{\mathbf{a}}_\perp \|\vec{\mathbf{a}}\|^2 \sin^2\alpha}{\|\vec{\mathbf{a}}_\perp\|^2}} =$$

$$= \sqrt{\|\vec{\mathbf{a}}\|^2 \cos^2\alpha + \|\vec{\mathbf{a}}\|^2 \sin^2\alpha} = \|\vec{\mathbf{a}}\|,$$

$$\angle(\vec{\mathbf{a}}, \vec{\mathbf{a}}') = \arccos\left(\frac{\vec{\mathbf{a}} \cdot \vec{\mathbf{a}}'}{\|\vec{\mathbf{a}}\|\|\vec{\mathbf{a}}'\|}\right) = \arccos\left(\frac{1}{\|\vec{\mathbf{a}}\|^2}\left(\vec{\mathbf{a}} \cdot \vec{\mathbf{a}} \cos\alpha + \frac{\vec{\mathbf{a}} \cdot \vec{\mathbf{a}}_\perp \|\vec{\mathbf{a}}\| \sin\alpha}{\|\vec{\mathbf{a}}_\perp\|}\right)\right) = \alpha.$$

As a linear combination of vectors $\vec{\mathbf{a}}$ and $\vec{\mathbf{b}}$, vector $\vec{\mathbf{a}}'$ lays in the same plane, and, moreover,

$$\vec{\mathbf{a}} \times \vec{\mathbf{a}}' = \vec{\mathbf{a}} \times \vec{\mathbf{a}} \cos\alpha + \vec{\mathbf{a}} \times \left(\vec{\mathbf{b}} - \frac{\vec{\mathbf{a}}\,(\vec{\mathbf{a}} \cdot \vec{\mathbf{b}})}{\vec{\mathbf{a}} \cdot \vec{\mathbf{a}}}\right) \frac{\|\vec{\mathbf{a}}\| \sin\alpha}{\|\vec{\mathbf{a}}_\perp\|} = \frac{\vec{\mathbf{a}} \times \vec{\mathbf{b}}\,\|\vec{\mathbf{a}}\| \sin\alpha}{\|\vec{\mathbf{a}}_\perp\|},$$

which implies that for $\alpha < \pi$ vector $\vec{\mathbf{a}}'$ belongs to the same half-plane as vector $\vec{\mathbf{b}}$ relative to the axis though vector $\vec{\mathbf{a}}$.

COROLLARY 3.38

The vector $\vec{\mathbf{a}}'' := \vec{\mathbf{a}} \overset{\alpha}{\vee} \vec{\mathbf{b}}$ is given by

$$\vec{\mathbf{a}}'' = \frac{\|\vec{\mathbf{a}}\|}{\|\vec{\mathbf{b}}\|} \vec{\mathbf{b}} \downarrow^{\vec{a}} \alpha = \left(\frac{\vec{\mathbf{b}} \cos\alpha}{\|\vec{\mathbf{b}}\|} + \frac{\vec{\mathbf{b}}_\perp \sin\alpha}{\|\vec{\mathbf{b}}_\perp\|}\right) \|\vec{\mathbf{a}}\|, \qquad (3.111)$$

where $\vec{\mathbf{b}}_\perp$ is defined according to equation (3.110) with $\vec{\mathbf{a}}$ and $\vec{\mathbf{b}}$ exchanged.

REMARK 3.39

$\vec{\mathbf{a}} \overset{\alpha}{\vee} \vec{\mathbf{b}}$ can be also computed as $\vec{\mathbf{a}} \downarrow^{\vec{b}} (\angle(\vec{\mathbf{a}}, \vec{\mathbf{b}}) - \alpha)$, but it is more efficient to use the expression (3.111), unless the angle between vectors $\vec{\mathbf{a}}$ and $\vec{\mathbf{b}}$ is already determined.

REMARK 3.40

$\vec{\mathbf{a}} \downarrow \alpha$ can be computed as $\vec{\mathbf{a}} \downarrow^{\vec{b}} \alpha$ with a random vector $\vec{\mathbf{b}}$, non-collinear to $\vec{\mathbf{a}}$.

DEFINITION 3.41

Let $\mathbf{M}_{\vec{v}}(\alpha)$ denote the matrix given by

$$\mathbf{M}_{\vec{v}}(\alpha) := \vec{\mathbf{v}}\vec{\mathbf{v}}^{\mathrm{T}} + (\mathbf{E} - \vec{\mathbf{v}}\vec{\mathbf{v}}^{\mathrm{T}}) \cos\alpha + \mathbf{S} \sin\alpha, \qquad (3.112)$$

where $\vec{\mathbf{v}} = (v^{[1]}, v^{[2]}, v^{[3]})^{\mathrm{T}}$ is a unit vector, $\mathbf{E} \in \mathbb{R}^{3\times 3}$ is a unit matrix and

$$\mathbf{S} := \begin{pmatrix} 0 & -v^{[3]} & v^{[2]} \\ v^{[3]} & 0 & -v^{[1]} \\ -v^{[2]} & v^{[1]} & 0 \end{pmatrix}.$$

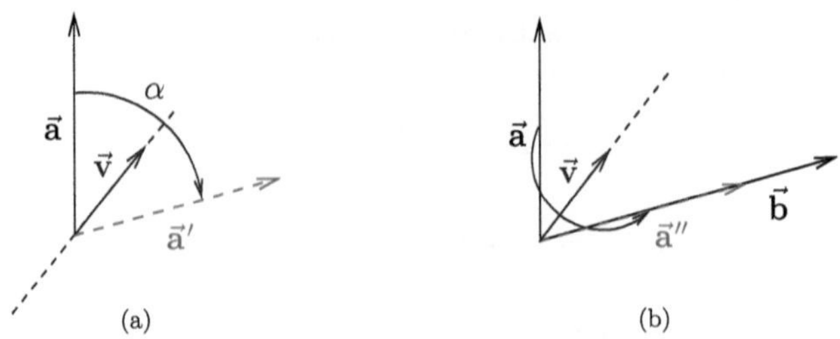

FIG. 3.13: Rotation about vector $\vec{v}$. (a) $\vec{a}' = \mathbf{M}_{\vec{v}}(\alpha)\vec{a}$, $\vec{v}$ points from the viewer. (b) $\vec{a}'' = \mathbf{M}_{\vec{b}}^{\vec{a}}\vec{a}$, $\vec{v}$ is between $\vec{a}$ and $\vec{b}$.

A multiplication of this matrix with a vector $\vec{a}$ results in *rotation** of $\vec{a}$ *to the angle* α *about vector* $\vec{v}$ [121]. The rotation is clockwise, when viewed along vector $\vec{v}$ (see Figure 3.13 (a)).

DEFINITION 3.42
Let $\mathbf{M}_{\vec{b}}^{\vec{a}}$ denote a rotation matrix, such that the vector $\mathbf{M}_{\vec{b}}^{\vec{a}}\vec{a}$ is directed as vector $\vec{b}$. We will say that multiplication of this matrix with vector $\vec{a}$ results in *matching the direction* of vector $\vec{a}$ with vector $\vec{b}$.

REMARK 3.43
The matrix for direction matching, which satisfies Definition 3.42, is not uniquely defined, since a subsequent rotation about vector $\vec{b}$ keeps the vector $\mathbf{M}_{\vec{b}}^{\vec{a}}\vec{a}$ unchanged.

One of possible realizations of this operation is to rotate vector $\vec{a}$ to the angle π about the unit vector

$$\vec{v} = \begin{pmatrix} v^{[1]} \\ v^{[2]} \\ v^{[3]} \end{pmatrix} = \frac{\frac{\vec{a}}{\|\vec{a}\|} + \frac{\vec{b}}{\|\vec{b}\|}}{\left\| \frac{\vec{a}}{\|\vec{a}\|} + \frac{\vec{b}}{\|\vec{b}\|} \right\|},$$

see Figure 3.13 (b).

Substitution of $\alpha = \pi$ into equation (3.112) gives:

$$\mathbf{M}_{\vec{b}}^{\vec{a}} = \begin{pmatrix} 2(v^{[1]})^2 - 1 & 2v^{[1]}v^{[2]} & 2v^{[1]}v^{[3]} \\ 2v^{[1]}v^{[2]} & 2(v^{[2]})^2 - 1 & 2v^{[2]}v^{[3]} \\ 2v^{[1]}v^{[3]} & 2v^{[2]}v^{[3]} & 2(v^{[3]})^2 - 1 \end{pmatrix}.$$

REMARK 3.44
Direction matching is utilized not for transforming the vector $\vec{a}$, which could be simply achieved by scaling the vector $\vec{b}$, but for rotating related objects.

---

*Again, it is implied that the original vector is not changed, but a new vector is created as a result of the operation.

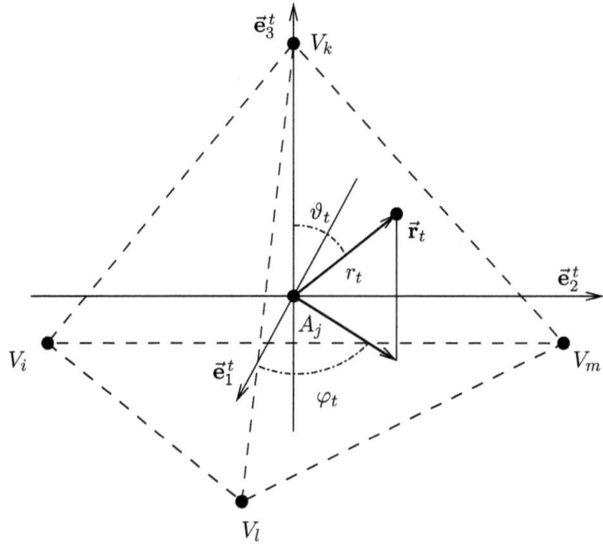

FIG. 3.14: A tetrahedron with related coordinate systems.

For generation of atomic coordinates by algorithms described in the next chapter, we need to solve the task of reconstructing the directions of two vertices of a regular tetrahedron relative to its center, when the center A_j and the directions $\vec{\mathbf{r}}_{jk}$ and $\vec{\mathbf{r}}_{jl}$ of two other vertices V_k, V_l are given. For this purpose we introduce a new local Cartesian coordinate system with the center in A_j and the orthonormal basis $\{\vec{\mathbf{e}}_1^t, \vec{\mathbf{e}}_2^t, \vec{\mathbf{e}}_3^t\}$, related to the tetrahedron so that

$$\vec{\mathbf{e}}_3^t =: \begin{pmatrix} t_3^{[1]} \\ t_3^{[2]} \\ t_3^{[3]} \end{pmatrix} := \frac{\vec{\mathbf{r}}_{jk}}{\|\vec{\mathbf{r}}_{jk}\|}, \tag{3.113}$$

$$\vec{\mathbf{e}}_2^t =: \begin{pmatrix} t_2^{[1]} \\ t_2^{[2]} \\ t_2^{[3]} \end{pmatrix} := \frac{\vec{\mathbf{r}}_{jk} \times \vec{\mathbf{r}}_{jl}}{\|\vec{\mathbf{r}}_{jk} \times \vec{\mathbf{r}}_{jl}\|}, \tag{3.114}$$

$$\vec{\mathbf{e}}_1^t =: \begin{pmatrix} t_1^{[1]} \\ t_1^{[2]} \\ t_1^{[3]} \end{pmatrix} := \vec{\mathbf{e}}_2^t \times \vec{\mathbf{e}}_3^t. \tag{3.115}$$

Knowing that the angle between vertex directions in a regular tetrahedron is equal to $\arccos(-1/3)$ (see Lemma 3.19), we can specify the positions of the left and the right vertex, V_i and V_m (see Figure 3.14), in the associated spherical coordinate system as $(r, -120°, 109.4712°)$ and $(r, 120°, 109.4712°)$, where r is the distance from any vertex to the center. Since we need only the directions, we set $r = 1$. The corresponding coordinates in the local Cartesian coordinate system can be then computed from spherical coordinates using the mapping given by Definition 3.27. Thus, we define:

$$\vec{\mathbf{v}}_l := \Upsilon(1, -120°, 109.4712°) \quad \text{and} \quad \vec{\mathbf{v}}_r := \Upsilon(1, 120°, 109.4712°). \tag{3.116}$$

The positions of vertices V_i and V_m in the original global Cartesian coordinate system can be reconstructed using $\vec{\mathbf{v}}_l$ and $\vec{\mathbf{v}}_r$ according to the following proposition. For another r vectors $\vec{\mathbf{v}}_l$ and $\vec{\mathbf{v}}_r$ can be scaled as appropriate.

PROPOSITION 3.45

Let $\vec{\mathbf{r}}_t$ be the position of a point in the local Cartesian coordinate system with the basis $\{\vec{\mathbf{e}}_1^t, \vec{\mathbf{e}}_2^t, \vec{\mathbf{e}}_3^t\}$ given by (3.113)-(3.115) and centered at the global position $\vec{\mathbf{r}}_j$. Then the global position $\vec{\mathbf{r}}$ of the considered point is given by

$$\vec{\mathbf{r}} = \vec{\mathbf{r}}_j + \mathbf{T}_{\vec{\mathbf{r}}_{jl}}^{\vec{\mathbf{r}}_{jk}} \vec{\mathbf{r}}_t,$$

where

$$\mathbf{T}_{\vec{\mathbf{r}}_{jl}}^{\vec{\mathbf{r}}_{jk}} := \begin{pmatrix} \vec{\mathbf{e}}_1^t & \vec{\mathbf{e}}_2^t & \vec{\mathbf{e}}_3^t \end{pmatrix} := \begin{pmatrix} t_1^{[1]} & t_2^{[1]} & t_3^{[1]} \\ t_1^{[2]} & t_2^{[2]} & t_3^{[2]} \\ t_1^{[3]} & t_2^{[3]} & t_3^{[3]} \end{pmatrix}. \tag{3.117}$$

Proof. For any position $\vec{\mathbf{r}}$ in the global coordinate system holds:

$$\vec{\mathbf{r}} = \vec{\mathbf{r}}_j + r_t^{[1]}\vec{\mathbf{e}}_1^t + r_t^{[2]}\vec{\mathbf{e}}_2^t + r_t^{[3]}\vec{\mathbf{e}}_3^t = \vec{\mathbf{r}}_j + \begin{pmatrix} r_t^{[1]}t_1^{[1]} + r_t^{[2]}t_2^{[1]} + r_t^{[3]}t_3^{[1]} \\ r_t^{[1]}t_1^{[2]} + r_t^{[2]}t_2^{[2]} + r_t^{[3]}t_3^{[2]} \\ r_t^{[1]}t_1^{[3]} + r_t^{[2]}t_2^{[3]} + r_t^{[3]}t_3^{[3]} \end{pmatrix} = \vec{\mathbf{r}}_j + \mathbf{T}_{\vec{\mathbf{r}}_{jl}}^{\vec{\mathbf{r}}_{jk}} \begin{pmatrix} r_t^{[1]} \\ r_t^{[2]} \\ r_t^{[3]} \end{pmatrix}. \quad \square$$

3.5.2　APPENDING A NEW AMINO ACID RESIDUE

Assume that amino acids with optimized atomic coordinates are available. Before we proceed, we shall take care that each amino group participating in peptide bond formation is transferred into the non-ionized form and, moreover, acquires an sp^2-like arrangement, typical for peptide groups.

Let $\mathcal{P}$ and $\mathcal{A}$ be three-dimensional Euclidian spaces equipped with Cartesian coordinate systems, and let $\vec{\mathbf{p}}_X \in \mathcal{P}$ and $\vec{\mathbf{a}}_X \in \mathcal{A}$ denote the vectors pointing from the chosen coordinate origin to the position of the object X, located in $\mathcal{P}$ or $\mathcal{A}$ respectively. Let the points H, N and C_α^a represent respectively one of the hydrogens of the amino group, the nitrogen and the α-carbon of the amino acid common part, and C', O, C_α^p be the centers of C', one of the oxygen atoms of the carboxyl group and the α-carbon of the protein current terminal residue. Naturally, we designate $\vec{\mathbf{p}}_{XY} := \vec{\mathbf{p}}_Y - \vec{\mathbf{p}}_X$ and $\vec{\mathbf{a}}_{XY} := \vec{\mathbf{a}}_Y - \vec{\mathbf{a}}_X$.

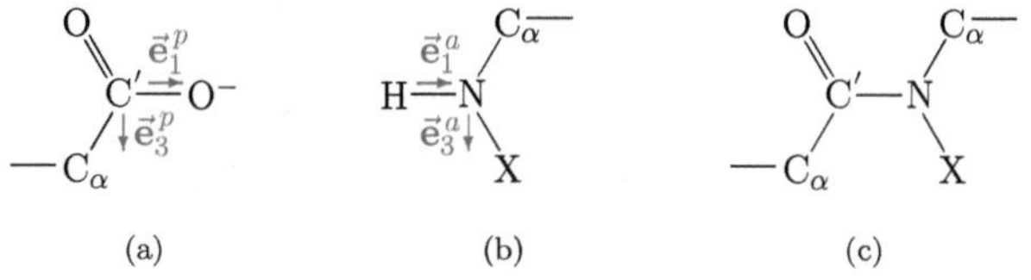

(a) (b) (c)

FIG. 3.15: Appending a new amino acid. Vectors $\vec{\mathbf{e}}_2^{\,p}$ and $\vec{\mathbf{e}}_2^{\,a}$ are directed to the viewer. X stays for C_γ in proline and for H in any other residue. Some bonds of C_α and C_γ are not shown.

Further, we introduce two auxiliary orthonormal bases $\vec{\mathbf{e}}_1^{\,a}$, $\vec{\mathbf{e}}_2^{\,a}$, $\vec{\mathbf{e}}_3^{\,a}$ and $\vec{\mathbf{e}}_1^{\,p}$, $\vec{\mathbf{e}}_2^{\,p}$, $\vec{\mathbf{e}}_3^{\,p}$, defined by equations (3.118)-(3.123).

$$\vec{\mathbf{e}}_1^{\,a} = \begin{pmatrix} a_1^{[1]} \\ a_1^{[2]} \\ a_1^{[3]} \end{pmatrix} := \frac{\vec{\mathbf{a}}_{HN}}{\|\vec{\mathbf{a}}_{HN}\|}, \tag{3.118}$$

$$\vec{\mathbf{e}}_2^{\,a} = \begin{pmatrix} a_2^{[1]} \\ a_2^{[2]} \\ a_2^{[3]} \end{pmatrix} := \frac{\vec{\mathbf{a}}_{HN} \times \vec{\mathbf{a}}_{NC_\alpha^a}}{\|\vec{\mathbf{a}}_{HN} \times \vec{\mathbf{a}}_{NC_\alpha^a}\|}, \tag{3.119}$$

$$\vec{\mathbf{e}}_3^{\,a} = \begin{pmatrix} a_3^{[1]} \\ a_3^{[2]} \\ a_3^{[3]} \end{pmatrix} := \vec{\mathbf{e}}_1^{\,a} \times \vec{\mathbf{e}}_2^{\,a}, \tag{3.120}$$

$$\vec{\mathbf{e}}_1^{\,p} = \begin{pmatrix} p_1^{[1]} \\ p_1^{[2]} \\ p_1^{[3]} \end{pmatrix} := \frac{\vec{\mathbf{p}}_{C'O}}{\|\vec{\mathbf{p}}_{C'O}\|}, \tag{3.121}$$

$$\vec{\mathbf{e}}_2^{\,p} = \begin{pmatrix} p_2^{[1]} \\ p_2^{[2]} \\ p_2^{[3]} \end{pmatrix} := \frac{\vec{\mathbf{p}}_{C'O} \times \vec{\mathbf{p}}_{C_\alpha^p C'}}{\|\vec{\mathbf{p}}_{C'O} \times \vec{\mathbf{p}}_{C_\alpha^p C'}\|}, \tag{3.122}$$

$$\vec{\mathbf{e}}_3^{\,p} = \begin{pmatrix} p_3^{[1]} \\ p_3^{[2]} \\ p_3^{[3]} \end{pmatrix} := \vec{\mathbf{e}}_1^{\,p} \times \vec{\mathbf{e}}_2^{\,p}. \tag{3.123}$$

Then the coordinate transformation, given in the following theorem, can be used for computation of new coordinates for the atoms of the appended amino acid residue.

THEOREM 3.46

Let l_p denote the reference length of the peptide bond and let $f : \mathcal{A} \to \mathcal{P}$ be an affine map such that

$$f(\vec{\mathbf{a}}) = \mathbf{P}\mathbf{A}^{\mathrm{T}}(\vec{\mathbf{a}} - \vec{\mathbf{a}}_N) + \vec{\mathbf{p}}_{C'} + l_p \frac{\vec{\mathbf{p}}_{C'O}}{\|\vec{\mathbf{p}}_{C'O}\|}, \quad \forall \vec{\mathbf{a}} \in \mathcal{A},$$

where $\mathbf{A}$ *and* $\mathbf{P}$ *are two matrices with columns representing the introduced auxiliary basis vectors:*

$$\mathbf{A} := \begin{pmatrix} \vec{\mathbf{e}}_1^a & \vec{\mathbf{e}}_2^a & \vec{\mathbf{e}}_3^a \end{pmatrix} := \begin{pmatrix} a_1^{[1]} & a_2^{[1]} & a_3^{[1]} \\ a_1^{[2]} & a_2^{[2]} & a_3^{[2]} \\ a_1^{[3]} & a_2^{[3]} & a_3^{[3]} \end{pmatrix},$$

$$\mathbf{P} := \begin{pmatrix} \vec{\mathbf{e}}_1^p & \vec{\mathbf{e}}_2^p & \vec{\mathbf{e}}_3^p \end{pmatrix} := \begin{pmatrix} p_1^{[1]} & p_2^{[1]} & p_3^{[1]} \\ p_1^{[2]} & p_2^{[2]} & p_3^{[2]} \\ p_1^{[3]} & p_2^{[3]} & p_3^{[3]} \end{pmatrix}.$$

If the new position $\vec{\mathbf{p}}$ *of each attached atom is computed from the old position* $\vec{\mathbf{a}}$ *as* $\vec{\mathbf{p}} = f(\vec{\mathbf{a}})$, *then the amino acid is appended in a trans conformation with all interatomic distances and chirality preserved, and the length of the peptide bond equal to* l_p.

Proof. At first we show that the geometry of the appended amino acid is retained.

$$\det \mathbf{A}^{\mathrm{T}} = \begin{vmatrix} (\vec{\mathbf{e}}_1^a)^{\mathrm{T}} \\ (\vec{\mathbf{e}}_2^a)^{\mathrm{T}} \\ (\vec{\mathbf{e}}_3^a)^{\mathrm{T}} \end{vmatrix} = (\vec{\mathbf{e}}_1^a \times \vec{\mathbf{e}}_2^a) \cdot \vec{\mathbf{e}}_3^a = \vec{\mathbf{e}}_3^a \cdot \vec{\mathbf{e}}_3^a = 1.$$

Reciprocally we obtain:

$$\det \mathbf{P} = \det \mathbf{P}^{\mathrm{T}} = 1, \ \det(\mathbf{P}\mathbf{A}^{\mathrm{T}}) = \det \mathbf{P} \det \mathbf{A}^{\mathrm{T}} = 1.$$

Hence $\mathbf{P}\mathbf{A}^{\mathrm{T}}$ is a rotation matrix. Consequently, the mapping f, as a superposition of rotation and translation, preserves distances and chirality.

The bond angles of the atoms C' and N are also maintained, since the directions of the vectors $\vec{\mathbf{p}}_{C'O}$, $\vec{\mathbf{p}}_{C'N}$ and $\vec{\mathbf{p}}_{HN}$ coincide:

$$\vec{\mathbf{p}}_{C'N} = f(\vec{\mathbf{a}}_N) - \vec{\mathbf{p}}_{C'} = \mathbf{P}\mathbf{A}^{\mathrm{T}}(\vec{\mathbf{a}}_N - \vec{\mathbf{a}}_N) + \vec{\mathbf{p}}_{C'} + l_p \frac{\vec{\mathbf{p}}_{C'O}}{\|\vec{\mathbf{p}}_{C'O}\|} - \vec{\mathbf{p}}_{C'} = l_p \frac{\vec{\mathbf{p}}_{C'O}}{\|\vec{\mathbf{p}}_{C'O}\|},$$

$$\vec{\mathbf{p}}_{HN} = f(\vec{\mathbf{a}}_N) - f(\vec{\mathbf{a}}_H) = -\mathbf{P}\mathbf{A}^{\mathrm{T}}(\vec{\mathbf{a}}_H - \vec{\mathbf{a}}_N) = \mathbf{P}\mathbf{A}^{\mathrm{T}}\vec{\mathbf{a}}_{HN} = \mathbf{P}\mathbf{A}^{\mathrm{T}}\vec{\mathbf{e}}_1^a \|\vec{\mathbf{a}}_{HN}\| =$$

$$= \begin{pmatrix} \vec{\mathbf{e}}_1^p & \vec{\mathbf{e}}_2^p & \vec{\mathbf{e}}_3^p \end{pmatrix} \begin{pmatrix} (\vec{\mathbf{e}}_1^a)^{\mathrm{T}} \\ (\vec{\mathbf{e}}_2^a)^{\mathrm{T}} \\ (\vec{\mathbf{e}}_3^a)^{\mathrm{T}} \end{pmatrix} \vec{\mathbf{e}}_1^a \|\vec{\mathbf{a}}_{HN}\| = \begin{pmatrix} \vec{\mathbf{e}}_1^p & \vec{\mathbf{e}}_2^p & \vec{\mathbf{e}}_3^p \end{pmatrix} \begin{pmatrix} 1 \\ 0 \\ 0 \end{pmatrix} \|\vec{\mathbf{a}}_{HN}\| =$$

$$= \vec{\mathbf{e}}_1^p \|\vec{\mathbf{a}}_{HN}\| = \frac{\vec{\mathbf{p}}_{C'O}}{\|\vec{\mathbf{p}}_{C'O}\|} \|\vec{\mathbf{a}}_{HN}\|.$$

The length of the formed peptide bond is

$$\|\vec{\mathbf{p}}_{C'N}\| = \left\| l_p \frac{\vec{\mathbf{p}}_{C'O}}{\|\vec{\mathbf{p}}_{C'O}\|} \right\| = l_p.$$

Now we prove that centers of the atoms C_α^p, C', N and C_α^a related to the created peptide group lay in one plane. $\vec{e}_2^p$ is orthogonal to $\vec{p}_{C_\alpha^p C'}$ and $\vec{p}_{C'N}$. Since the latter two vectors are non-collinear, it is enough to show that

$$
\vec{e}_2^p \cdot \vec{p}_{NC_\alpha^a} = \vec{e}_2^p \cdot (\mathbf{PA}^{\mathrm{T}} \vec{a}_{NC_\alpha^a}) = (\vec{e}_2^p)^{\mathrm{T}} \begin{pmatrix} \vec{e}_1^p & \vec{e}_2^p & \vec{e}_3^p \end{pmatrix} \begin{pmatrix} (\vec{e}_1^a)^{\mathrm{T}} \\ (\vec{e}_2^a)^{\mathrm{T}} \\ (\vec{e}_3^a)^{\mathrm{T}} \end{pmatrix} \vec{a}_{NC_\alpha^a} =
$$

$$
= \begin{pmatrix} 0 & 1 & 0 \end{pmatrix} \begin{pmatrix} (\vec{e}_1^a)^{\mathrm{T}} \\ (\vec{e}_2^a)^{\mathrm{T}} \\ (\vec{e}_3^a)^{\mathrm{T}} \end{pmatrix} \vec{a}_{NC_\alpha^a} = (\vec{e}_2^a)^{\mathrm{T}} \vec{a}_{NC_\alpha^a} = \vec{e}_2^a \cdot \vec{a}_{NC_\alpha^a} = 0.
$$

Let $P_\perp$ be the plane that passes through the centers of atoms C' and N perpendicular to the plane of the peptide group, and let $\alpha, \beta \in (0, \pi)$ be the angles confined between vectors $\vec{p}_{C'O}$, $\vec{p}_{C_\alpha^p C'}$ and $\vec{a}_{HN}$, $\vec{a}_{NC_\alpha^a}$ respectively*. On construction, $\vec{e}_3^p$ is a normal to $P_\perp$. We demonstrate that the atoms C_α^p and C_α^a will be situated on different sides of $P_\perp$:

$$
\vec{p}_{C'C_\alpha^p} \cdot \vec{e}_3^p = \vec{p}_{C'C_\alpha^p} \cdot (\vec{e}_1^p \times \vec{e}_2^p) = \vec{p}_{C'C_\alpha^p} \cdot \left(\vec{e}_1^p \times \frac{(\vec{p}_{C'O} \times \vec{p}_{C_\alpha^p C'})}{\|\vec{p}_{C'O} \times \vec{p}_{C_\alpha^p C'}\|} \right) =
$$

$$
= -\frac{\vec{p}_{C_\alpha^p C'}}{\|\vec{p}_{C'O} \times \vec{p}_{C_\alpha^p C'}\|} \cdot \left(\vec{e}_1^p \times (\vec{p}_{C'O} \times \vec{p}_{C_\alpha^p C'}) \right) =
$$

$$
= -\frac{\vec{p}_{C_\alpha^p C'}}{\|\vec{p}_{C'O} \times \vec{p}_{C_\alpha^p C'}\|} \cdot \left(\vec{p}_{C'O}(\vec{e}_1^p \cdot \vec{p}_{C_\alpha^p C'}) - \vec{p}_{C_\alpha^p C'}(\vec{e}_1^p \cdot \vec{p}_{C'O}) \right) =
$$

$$
= -\frac{1}{\|\vec{p}_{C'O}\| \|\vec{p}_{C_\alpha^p C'}\| \sin \alpha} \left(\|\vec{p}_{C_\alpha^p C'}\|^2 \|\vec{p}_{C'O}\| \cos^2 \alpha - \|\vec{p}_{C_\alpha^p C'}\|^2 \|\vec{p}_{C'O}\| \right) =
$$

$$
= \|\vec{p}_{C_\alpha^p C'}\| \sin \alpha > 0,
$$

$$
\vec{p}_{NC_\alpha^a} \cdot \vec{e}_3^p = \vec{e}_3^p \cdot (\mathbf{PA}^{\mathrm{T}} \vec{a}_{NC_\alpha^a}) = (\vec{e}_3^p)^{\mathrm{T}} \begin{pmatrix} \vec{e}_1^p & \vec{e}_2^p & \vec{e}_3^p \end{pmatrix} \begin{pmatrix} (\vec{e}_1^a)^{\mathrm{T}} \\ (\vec{e}_2^a)^{\mathrm{T}} \\ (\vec{e}_3^a)^{\mathrm{T}} \end{pmatrix} \vec{a}_{NC_\alpha^a} =
$$

$$
= \begin{pmatrix} 0 & 0 & 1 \end{pmatrix} \begin{pmatrix} (\vec{e}_1^a)^{\mathrm{T}} \\ (\vec{e}_2^a)^{\mathrm{T}} \\ (\vec{e}_3^a)^{\mathrm{T}} \end{pmatrix} \vec{a}_{NC_\alpha^a} = \vec{e}_3^a \cdot \vec{a}_{NC_\alpha^a} =
$$

$$
= \frac{\vec{a}_{NC_\alpha^a}}{\|\vec{a}_{HN} \times \vec{a}_{NC_\alpha^a}\|} \cdot \left(\vec{e}_1^a \times (\vec{a}_{HN} \times \vec{a}_{NC_\alpha^a}) \right) =
$$

$$
= \frac{\vec{a}_{NC_\alpha^a}}{\|\vec{a}_{HN} \times \vec{a}_{NC_\alpha^a}\|} \cdot \left(\vec{a}_{HN}(\vec{e}_1^a \cdot \vec{a}_{NC_\alpha^a}) - \vec{a}_{NC_\alpha^a}(\vec{e}_1^a \cdot \vec{a}_{HN}) \right) =
$$

$$
= \frac{1}{\|\vec{a}_{HN}\| \|\vec{a}_{NC_\alpha^a}\| \sin \beta} \left(\|\vec{a}_{NC_\alpha^a}\|^2 \|\vec{a}_{HN}\| \cos^2 \beta - \|\vec{a}_{NC_\alpha^a}\|^2 \|\vec{a}_{HN}\| \right) =
$$

$$
= -\|\vec{a}_{NC_\alpha^a}\| \sin \beta < 0.
$$

---

*Using reference bond angles one can show that $\alpha \approx \beta \approx \frac{\pi}{3}$.

Hence the created peptide group will be in *trans* conformation. $\square$

COROLLARY 3.47

We get the cis conformation with preserved chirality if we change the directions of $\vec{e}_2^{a}$ and $\vec{e}_3^{a}$, or $\vec{e}_2^{p}$ and $\vec{e}_3^{p}$, to the opposite.

3.6 TWISTING FORCES

Before proceeding to the discussion of twisting forces related to each degree of freedom, we shall consider forces acting on single atoms.

DEFINITION 3.48

Let the vector of partial derivatives of a function $f(\vec{r}_1, \ldots, \vec{r}_N)$ with respect to the coordinates of an atom position $\vec{r}_i$ be denoted as $\partial f/\partial \vec{r}_i$:

$$\frac{\partial f}{\partial \vec{r}_i} := \left(\frac{\partial f}{\partial r_i^{[1]}}, \frac{\partial f}{\partial r_i^{[2]}}, \frac{\partial f}{\partial r_i^{[3]}} \right)^{\mathrm{T}}.$$

REMARK 3.49

Note that for a given energy function $U(\vec{r}_1, \ldots, \vec{r}_N)$, the derivative $\partial U/\partial \vec{r}_i$ taken with the opposite sign corresponds to the force acting on atom A_i.

For computation of forces arising from different interactions we shall need the following lemma.

LEMMA 3.50

For any $i, j \in \mathcal{N}$ holds:

$$\frac{\partial \|\vec{r}_{ij}\|}{\partial \vec{r}_i} = \vec{e}_{ji}.$$

Proof. For $l = \overline{1,3}$ we have:

$$\frac{\partial \|\vec{r}_{ij}\|}{\partial r_i^{[l]}} = \frac{\partial \sqrt{\sum_{k=1}^{3} \left(r_j^{[k]} - r_i^{[k]} \right)^2}}{\partial r_i^{[l]}} = \frac{r_i^{[l]} - r_j^{[l]}}{\sqrt{\sum_{k=1}^{3} \left(r_j^{[k]} - r_i^{[k]} \right)^2}} = e_{ji}^{[l]}. \quad \square$$

Lemma 3.50 can be used, for example, for computation of forces arising from bond stretching. Assume that we want to allow formation of disulfide bridges when two thiol groups approach each other. Although the bond lengths and angles are fixed in the suggested model, it can make sense to supplement the force field by terms describing bond stretching and angle bending related to formed disulfide bonds, to enable their breakage at certain conditions and new formation.

Let for two sulfur atoms A_i and A_j participating in formation of a disulfide bridge, the bond stretching energy term be given by

$$U_{ij} = k_{SS}^{[b]}(\|\vec{\mathbf{r}}_{ij}\| - l_{SS})^2, \tag{3.124}$$

where $k_{SS}^{[b]}$ and l_{SS} are appropriate parameters, and let the angle bending term related to all bond angles in any sulfur atom A_j participating in disulfide bonding be given by

$$U_j^{[a]} = \sum_{\substack{i,k\in\mathcal{N} \\ i\bowtie j\bowtie k}} k_s^{[a]}(\angle(\vec{\mathbf{r}}_{ji}, \vec{\mathbf{r}}_{jk}) - \alpha_s)^2, \tag{3.125}$$

with parameters $k_s^{[a]}$ and α_S (cf. the first and the second term on the right-hand side of (2.2)). Since sulfur has only two bonds, the latter sum reduces to only one term.

Let $\mathcal{N}_{SS}$ be the set of atom numbers corresponding to sulfur atoms currently participating in disulfide bonding. Then the bond stretching force $\vec{\mathbf{f}}_i^{[b]}$ acting on A_i ($i \in \mathcal{N}_{SS}$) is given by

$$\vec{\mathbf{f}}_i^{[b]} = -\frac{\partial U_{ij}^{[b]}}{\partial \vec{\mathbf{r}}_i} = -2k_{SS}^{[b]}(\|\vec{\mathbf{r}}_{ij}\| - l_{SS})\vec{\mathbf{e}}_{ji}, \quad (j \in \mathcal{N}_{SS}) \wedge (j \bowtie i),$$

according to Remark 3.49 and Lemma 3.50.

Besides holds:

$$\frac{\partial(\vec{\mathbf{r}}_{ji} \cdot \vec{\mathbf{r}}_{jk})}{\partial \vec{\mathbf{r}}_i} = \vec{\mathbf{r}}_{jk},$$

$$\frac{\partial(\vec{\mathbf{r}}_{ji} \cdot \vec{\mathbf{r}}_{jk})}{\partial \vec{\mathbf{r}}_j} = \vec{\mathbf{r}}_{ij} + \vec{\mathbf{r}}_{kj},$$

since for all $l = \overline{1,3}$ we have:

$$\frac{\partial(\vec{\mathbf{r}}_{ji} \cdot \vec{\mathbf{r}}_{jk})}{\partial r_i^{[l]}} = \frac{\partial\left(\sum_{n=1}^3 (r_i^{[n]} - r_j^{[n]})r_{jk}^{[n]}\right)}{\partial r_i^{[l]}} = r_{jk}^{[l]},$$

$$\frac{\partial(\vec{\mathbf{r}}_{ji} \cdot \vec{\mathbf{r}}_{jk})}{\partial r_j^{[l]}} = \frac{\partial\left(\sum_{n=1}^3 (r_i^{[n]} - r_j^{[n]})(r_k^{[n]} - r_j^{[n]})\right)}{\partial r_j^{[l]}} = -r_i^{[l]} - r_k^{[l]} + 2r_j^{[l]} = r_{ij}^{[l]} + r_{kj}^{[l]}.$$

Therefore, the angle bending force $\vec{\mathbf{f}}_i^{[a]}$ acting on atom A_i is given for $i \in \mathcal{N}_{SS}$ by

$$\vec{\mathbf{f}}_i^{[a]} = -\frac{\partial\left(U_i^{[a]} + U_j^{[a]}\right)}{\partial \vec{\mathbf{r}}_i}, \quad (j \in \mathcal{N}_{SS}) \wedge (j \bowtie i),$$

and for $i \in \{n \in \mathcal{N} \setminus \mathcal{N}_{SS} \mid (n \bowtie m) \wedge (m \in \mathcal{N}_{SS})\}$ by

$$\vec{\mathbf{f}}_i^{[a]} = -\frac{\partial U_j^{[a]}}{\partial \vec{\mathbf{r}}_i}, \quad (j \in \mathcal{N}_{SS}) \wedge (j \bowtie i),$$

where

$$\frac{\partial U_i^{[a]}}{\partial \vec{\mathbf{r}}_i} = 2k_S^{[a]} \left(\arccos\left(\frac{\vec{\mathbf{r}}_{ij} \cdot \vec{\mathbf{r}}_{ik}}{\|\vec{\mathbf{r}}_{ij}\|\|\vec{\mathbf{r}}_{ik}\|} \right) - \alpha_S \right) \times$$

$$\times \frac{-1}{\sqrt{1 - \left(\frac{\vec{\mathbf{r}}_{ij} \cdot \vec{\mathbf{r}}_{ik}}{\|\vec{\mathbf{r}}_{ij}\|\|\vec{\mathbf{r}}_{ik}\|} \right)^2}} \frac{(\vec{\mathbf{r}}_{ij} \cdot \vec{\mathbf{r}}_{ik})(\vec{\mathbf{r}}_{ij} + \vec{\mathbf{r}}_{ik}) - \vec{\mathbf{r}}_{ij} - \vec{\mathbf{r}}_{ik}}{\|\vec{\mathbf{r}}_{ij}\|\|\vec{\mathbf{r}}_{ik}\|}, \quad (k \in \mathcal{N} \backslash \mathcal{N}_{SS}) \wedge (j \bowtie i \bowtie k),$$

and

$$\frac{\partial U_j^{[a]}}{\partial \vec{\mathbf{r}}_i} = 2k_S^{[a]} \left(\arccos\left(\frac{\vec{\mathbf{r}}_{ji} \cdot \vec{\mathbf{r}}_{jk}}{\|\vec{\mathbf{r}}_{ji}\|\|\vec{\mathbf{r}}_{jk}\|} \right) - \alpha_S \right) \times$$

$$\times \frac{-1}{\sqrt{1 - \left(\frac{\vec{\mathbf{r}}_{ji} \cdot \vec{\mathbf{r}}_{jk}}{\|\vec{\mathbf{r}}_{ji}\|\|\vec{\mathbf{r}}_{jk}\|} \right)^2}} \frac{\vec{\mathbf{r}}_{jk} - (\vec{\mathbf{r}}_{ji} \cdot \vec{\mathbf{r}}_{jk})\vec{\mathbf{r}}_{ji}}{\|\vec{\mathbf{r}}_{ji}\|\|\vec{\mathbf{r}}_{jk}\|}, \quad (k \in \mathcal{N}) \wedge (i \bowtie j \bowtie k).$$

Other forces typically can be computed in the same way. However, the hydration model described in Section 3.4 does not permit computation of solvation forces. Instead, the forces arising due to interaction with water can be implemented as semi-random forces acting from the side of solvent-accessible points on the atomic solvation grid (or in the direction of the exposed points, depending on the desired effect). Thus, if a hydration of a certain atom is unfavorable, forces must push it away from the surface. Otherwise, attractive forces from the side of water should dominate.

Consider now $\widetilde{U}^{[w]}(\vec{\zeta})$ and $\widetilde{G}^{[e]}(\vec{\zeta})$ given by equations (3.88) and (3.89). First of all, we have:

$$\widetilde{U}^{[w]}(\vec{\zeta}) = \sum_{\substack{(i,j)\in\mathcal{C} \\ d_{ij}(\vec{\zeta}) < C^{[w]}}} \left(U_{ij}^{[w]} \circ d_{ij} \right)(\vec{\zeta})$$

with

$$U_{ij}^{[w]}(d_{ij}) = \sqrt{E_i^{[w]} E_j^{[w]}} \left(\left(\frac{R_i^{[w]} + R_j^{[w]}}{d_{ij}} \right)^{12} - 2 \left(\frac{R_i^{[w]} + R_j^{[w]}}{d_{ij}} \right)^{6} \right), \quad \forall (i,j) \in \mathcal{C}.$$

Hence,

$$\frac{\partial \widetilde{U}^{[w]}(\vec{\zeta})}{\partial \zeta^{[k]}} = \sum_{\substack{(i,j)\in\mathcal{C} \\ d_{ij}(\vec{\zeta}) < C^{[w]}}} \left(U_{ij}^{[w]} \right)'(d_{ij}) \frac{\partial d_{ij}(\vec{\zeta})}{\partial \zeta^{[k]}},$$

where

$$\left(U_{ij}^{[w]} \right)'(d_{ij}) = -\frac{12}{d_{ij}} \sqrt{E_i^{[w]} E_j^{[w]}} \left(\left(\frac{R_i^{[w]} + R_j^{[w]}}{d_{ij}} \right)^{12} - \left(\frac{R_i^{[w]} + R_j^{[w]}}{d_{ij}} \right)^{6} \right).$$

However, instead of differentiating (3.90), it is more simple to do the following computations:

$$\frac{\partial d_{ij}(\vec{\zeta})}{\partial \zeta^{[k]}} = \sum_{l=1}^{3} \left(\frac{\partial d_{ij}}{\partial r_i^{[l]}} \frac{\partial r_i^{[l]}}{\partial \zeta^{[k]}} + \frac{\partial d_{ij}}{\partial r_j^{[l]}} \frac{\partial r_j^{[l]}}{\partial \zeta^{[k]}} \right) = \frac{\partial d_{ij}(\vec{\zeta})}{\partial \vec{r}_i} \cdot \frac{\partial \vec{r}_i}{\partial \zeta^{[k]}} + \frac{\partial d_{ij}(\vec{\zeta})}{\partial \vec{r}_j} \cdot \frac{\partial \vec{r}_j}{\partial \zeta^{[k]}}, \quad (3.126)$$

noting that $d_{ij} = \|\vec{r}_{ij}\|$. By Lemma 3.50,

$$\frac{\partial d_{ij}}{\partial \vec{r}_i} = \vec{e}_{ji} \quad\quad (3.127)$$

and, in fact,

$$-\left(U_{ij}^{[w]} \right)'(d_{ij}) \, \vec{e}_{ji} = \left(U_{ij}^{[w]} \right)'(d_{ij}) \, \vec{e}_{ij}$$

is the force acting on atom A_i due to the van der Waals interaction between atoms A_i and A_j (see Remark 3.49).

As for $\widetilde{G}^{[e]}(\vec{\zeta})$, there is an additional complication arising from the dependence of the screening functions on dihedral angles. However, if we assume that ϵ_{ij}, $\forall (i,j) \in \mathcal{C}$, mostly remain constant for small changes of distances between atoms, we can similarly deduce that

$$-\left(G_{ij}^{[e]} \right)'(d_{ij}) \, \vec{e}_{ji}$$

is the electrostatic force acting on atom A_i due to the interaction with atom A_j, where

$$\left(G_{ij}^{[e]} \right)'(d_{ij}) \approx -\frac{q_i q_j}{4\pi \epsilon_0 \epsilon_{ij} d_{ij}^2}.$$

The computation of $\partial \vec{r}_i / \partial \zeta^{[k]}$ can be done according to Proposition 3.51. A more general discussion about twisting forces follows in Proposition 3.52.

PROPOSITION 3.51
Let $\vec{\zeta}$ be a vector of primary dihedral angles completely determining the conformation of the molecule, and let any change in dihedral angle $\zeta^{[i]}$, $i \in \mathcal{M}$, be always achieved by rotation of the shortest (if applicable) branch $\mathcal{D}_{jk}$, such that $i = \kappa(j,k)$. To be more specific, let

$$\mathcal{X}_{1/2} = \{ (i,j) \in \mathcal{X} \mid (w_{ij} < w_{ji}) \vee ((w_{ij} = w_{ji}) \wedge (i < j)) \}$$

(cf. 3.85), and let $\iota : \mathcal{M} \to \mathcal{X}_{1/2}$ be the map that relates the indices of degrees of freedom with indices of the corresponding branches.

Let $i \in \mathcal{M}$, $(j,k) = \iota(i)$, and $m \in \mathcal{N}$. Then holds:

$$\frac{\partial \vec{r}_m}{\partial \zeta^{[i]}} = \begin{cases} \vec{e}_{jk} \times \vec{r}_{km}, & \text{for } m \in \mathcal{N}_{jk}, \\ 0 & \text{otherwise.} \end{cases} \quad\quad (3.128)$$

Proof. When dihedral angle $\zeta^{[i]}$ is increased by value $\Delta\zeta$, the position of atom A_m is changed according to (3.129)*, if $m \in \mathcal{N}_{jk}$:

$$\vec{\mathbf{r}}_m(\zeta^{[1]},\ldots,\zeta^{[i-1]},\zeta^{[i]}+\Delta\zeta,\zeta^{[i+1]},\ldots,\zeta^{[M]}) = \vec{\mathbf{r}}_k(\vec{\zeta}) + \mathbf{M}_{\vec{e}_{jk}}(\Delta\zeta)\left(\vec{\mathbf{r}}_m(\vec{\zeta}) - \vec{\mathbf{r}}_k(\vec{\zeta})\right),$$

$$\tag{3.129}$$

and does not change if $m \in \mathcal{N}_{kj}$.

Hence, for $m \in \mathcal{N}_{jk}$ holds:

$$\frac{\partial\vec{\mathbf{r}}_m}{\partial\zeta^{[i]}} = \lim_{\Delta\zeta\to 0}\frac{\vec{\mathbf{r}}_k(\vec{\zeta}) + \mathbf{M}_{\vec{e}_{jk}}(\Delta\zeta)\left(\vec{\mathbf{r}}_m(\vec{\zeta}) - \vec{\mathbf{r}}_k(\vec{\zeta})\right) - \vec{\mathbf{r}}_m(\vec{\zeta})}{\Delta\zeta} =$$

$$= \lim_{\Delta\zeta\to 0}\frac{\mathbf{M}_{\vec{e}_{jk}}(\Delta\zeta) - \mathbf{E}}{\Delta\zeta}\left(\vec{\mathbf{r}}_m(\vec{\zeta}) - \vec{\mathbf{r}}_k(\vec{\zeta})\right) = \mathbf{M}'_{\vec{e}_{jk}}(0)\,\vec{\mathbf{r}}_{km}(\vec{\zeta}), \tag{3.130}$$

and for $m \in \mathcal{N}_{kj}$ we have:

$$\frac{\partial\vec{\mathbf{r}}_m}{\partial\zeta^{[i]}} = 0. \tag{3.131}$$

From (3.112) we obtain:

$$\mathbf{M}'_{\vec{e}_{jk}}(0) = \begin{pmatrix} 0 & -e_{jk}^{[3]} & e_{jk}^{[2]} \\ e_{jk}^{[3]} & 0 & -e_{jk}^{[1]} \\ -e_{jk}^{[2]} & e_{jk}^{[1]} & 0 \end{pmatrix},$$

and note that

$$\mathbf{M}'_{\vec{e}_{jk}}(0)\,\vec{\mathbf{r}} = \vec{\mathbf{e}}_{jk} \times \vec{\mathbf{r}}, \quad \vec{\mathbf{r}} \in \mathbb{R}^3. \;\square \tag{3.132}$$

Substitution of (3.132) into (3.130) together with 3.131 yields 3.128. $\square$

PROPOSITION 3.52
Let

$$\mathcal{N}_{\nsim m} := \{n \in \mathcal{N} \mid n \nsim m\},$$

and let $\vec{\mathbf{f}}_m$ be the sum of all forces[†] acting on atom A_m:

$$\vec{\mathbf{f}}_m := -\sum_{n\in\mathcal{N}_{\nsim m}}\frac{\partial\widetilde{U}}{\partial d_{nm}}\,\vec{\mathbf{e}}_{nm}.$$

Under the conditions of Proposition 3.51 holds:

$$\frac{\partial\widetilde{U}(\vec{\zeta})}{\partial\zeta^{[i]}} = -\sum_{m\in\mathcal{N}_{jk}}\vec{\mathbf{f}}_m\cdot(\vec{\mathbf{e}}_{jk}\times\vec{\mathbf{r}}_{km}(\vec{\zeta})). \tag{3.133}$$

---

*The rotation matrix $\mathbf{M}_{\vec{e}_{jk}}(\Delta\zeta)$ is given by (3.112) with appropriate substitutions.

[†]van der Waals forces for distances above the cutoff are considered to be negligible.

Proof. According to (3.87), (3.126), and (3.127) we have:

$$\frac{\partial \widetilde{U}(\vec{\zeta})}{\partial \zeta^{[i]}} = \sum_{(m,n)\in\mathcal{C}} \frac{\partial \widetilde{U}}{\partial d_{mn}} \left(\frac{\partial d_{mn}}{\partial \vec{\mathbf{r}}_m} \cdot \frac{\partial \vec{\mathbf{r}}_m}{\partial \zeta^{[i]}} + \frac{\partial d_{mn}}{\partial \vec{\mathbf{r}}_n} \cdot \frac{\partial \vec{\mathbf{r}}_n}{\partial \zeta^{[i]}} \right) =$$

$$= \sum_{m\in\mathcal{N}} \sum_{n\in\mathcal{N}_{\approx m}} \frac{\partial \widetilde{U}}{\partial d_{nm}} \frac{\partial d_{nm}}{\partial \vec{\mathbf{r}}_m} \cdot \frac{\partial \vec{\mathbf{r}}_m}{\partial \zeta^{[i]}} =$$

$$= \sum_{m\in\mathcal{N}} \sum_{n\in\mathcal{N}_{\approx m}} \frac{\partial \widetilde{U}}{\partial d_{nm}} \vec{\mathbf{e}}_{nm} \cdot \frac{\partial \vec{\mathbf{r}}_m}{\partial \zeta^{[i]}} = -\sum_{m\in\mathcal{N}} \vec{\mathbf{f}}_m \cdot \frac{\partial \vec{\mathbf{r}}_m}{\partial \zeta^{[i]}}. \quad (3.134)$$

Substitution of (3.128) into (3.134) gives (3.133). $\square$

REMARK 3.53

If $m, n \in \mathcal{N}_{jk}$, then

$$\frac{\partial \widetilde{U}}{\partial d_{mn}} \left(\vec{\mathbf{e}}_{nm} \cdot \frac{\partial \vec{\mathbf{r}}_m}{\partial \zeta^{[i]}} + \vec{\mathbf{e}}_{mn} \cdot \frac{\partial \vec{\mathbf{r}}_n}{\partial \zeta^{[i]}} \right) =$$

$$= \frac{\partial \widetilde{U}}{\partial d_{mn}} \vec{\mathbf{e}}_{nm} \cdot \frac{\partial \vec{\mathbf{r}}_{nm}}{\partial \zeta^{[i]}} = \frac{\partial \widetilde{U}}{\partial d_{mn}} \vec{\mathbf{e}}_{nm} \cdot (\vec{\mathbf{e}}_{jk} \times \vec{\mathbf{r}}_{nm}) = 0,$$

therefore, it is sufficient to compute forces only between atoms of mutually complementary branches. Let

$$\mathcal{G}_{jk} := \{(m,n) \in \mathcal{N}^2 \mid (m \in \mathcal{N}_{jk}) \wedge (n \in \mathcal{N}_{kj})\}.$$

Then

$$\frac{\partial \widetilde{U}(\vec{\zeta})}{\partial \zeta^{[i]}} = -\sum_{(m,n)\in\mathcal{G}_{jk}} \vec{\mathbf{f}}_{nm} \cdot (\vec{\mathbf{e}}_{jk} \times \vec{\mathbf{r}}_{km}(\vec{\zeta})),$$

where

$$\vec{\mathbf{f}}_{nm} := -\frac{\partial \widetilde{U}}{\partial d_{nm}} \vec{\mathbf{e}}_{nm}$$

is the force acting on atom A_m due to interaction with atom A_n.

However, for computation of the complete energy gradient, it is technically more efficient to use (3.133) with preliminary computation of the sum of all forces acting on any atom.

From (3.133) we see that forces acting on atoms near a tip of a long branch can[*] make larger contribution to the energy gradient then those near the branch origin, by analogy to the principle of lever functioning. Therefore, the energy minimization problem becomes very stiff for sufficiently long chains. Thus, if the steepest descent in dihedral angle space is used for energy minimization, the rotations of

---

[*]Depending on the angles between the vectors in (3.133).

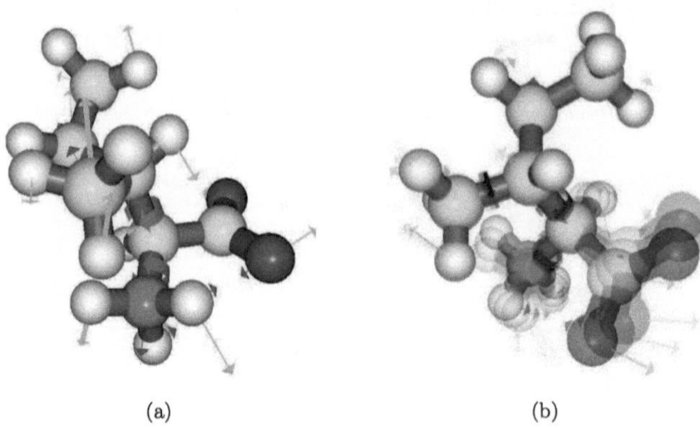

(a)   (b)

Fig. 3.16: Intramolecular atomic and twisting forces acting on a dissolved isoleucine molecule (a result of simulation by SiViProF, see Chapter 4): (a) a high-energy configuration showing strong van der Waals repulsion, (b) rotation of branches (depicted by motion blur) according to steepest descent dictated by twisting forces. See further explanation in the text.

long branches are very likely to dominate. However, it is natural to expect that in reality short branches rotate more easily. A more detailed discussion concerning this issue can be found in the next section.

We shall refer to $\partial \widetilde{U}(\vec{\zeta})/\partial \zeta^{[i]}$ as the *twisting force* acting on the bond between atoms A_j and A_k (such that $\kappa(j, k) = i$). Figure 3.16 shows computed atomic van der Waals and electrostatic forces together with the corresponding twisting forces acting to rotate branches.

In this figure and further, straight red and yellow arrows denote respectively the electrostatic and van der Waals forces acting on atoms. The absolute value of a force is shown by the arrow length* for forces up to $1 \, \text{kcal} \times (\text{mol} \times \text{Å})^{-1}$. The magnitudes of forces exceeding $1 \, \text{kcal} \times (\text{mol} \times \text{Å})^{-1}$ are depicted by respectively increased arrow width (see Figure 3.16 (a)). Curved arrows denote twisting forces. These arrows are shifted to one or another end of a bond, in order to show, which of the related branches is to be rotated. The absolute value of a twisting force up to 10 kcal/mol (subject to settings) is shown by the arrow width. In this case, the color of the arrow is dark red or blue, depending on whether the related primary dihedral angle must be increased or decreased (see Figure 3.16 (b)). If the absolute value of a twisting force exceeds the maximal value that can be indicated by the arrow width, the difference between the two mentioned magnitudes, up to a certain value (here also 10 kcal/mol, subject to an additional setting), is coded by the intensity of green, which is mixed into the original arrow color (as in Figure 3.16 (a)). The maximal green intensity is used for all values above the prescribed limit.

---

*The length is counted starting from the depicted atom surface.

3.7 Dynamics in dihedral angle space

The Lagrange function (see, for example, [122]) for an isolated system of N point masses, $m_1, \ldots, m_N$, with M degrees of freedom, $\zeta^{[1]}, \ldots, \zeta^{[M]}$, is given by

$$L(\zeta^{[1]}, \ldots, \zeta^{[M]}, \dot{\zeta}^{[1]}, \ldots, \dot{\zeta}^{[M]}) = \frac{1}{2} \sum_{k=1}^{N} m_k \|\dot{\vec{r}}_k\|^2 - U(\vec{r}_1, \ldots, \vec{r}_N), \qquad (3.135)$$

where

$$\vec{r}_k = \vec{r}_k(\zeta^{[1]}, \ldots, \zeta^{[M]}), \quad k = \overline{1, N}$$

are the positions of the point masses and

$$U(\vec{r}_1, \ldots, \vec{r}_N) = \widetilde{U}(\zeta^{[1]}, \ldots, \zeta^{[M]})$$

is the potential energy of the system.

We shall use it to describe the constrained dynamics of atoms A_k with masses m_k $(k = \overline{1, N})$ in vacuum. Then we supplement the obtained system of Lagrange equations,

$$\frac{d}{dt} \frac{\partial L}{\partial \dot{\zeta}^{[i]}} - \frac{\partial L}{\partial \zeta^{[i]}} = 0, \quad i = \overline{1, M}, \qquad (3.136)$$

to account for drag forces due to interaction with the solvent.

First of all, we have:

$$\dot{\vec{r}}_k = \sum_{j=1}^{M} \frac{\partial \vec{r}_k}{\partial \zeta^{[j]}} \dot{\zeta}^{[j]}, \qquad (3.137)$$

and hence

$$\frac{\partial(\dot{\vec{r}}_k \cdot \dot{\vec{r}}_k)}{\partial \dot{\zeta}^{[i]}} = 2\dot{\vec{r}}_k \cdot \frac{\partial \dot{\vec{r}}_k}{\partial \dot{\zeta}^{[i]}} = 2 \left(\sum_{j=1}^{M} \frac{\partial \vec{r}_k}{\partial \zeta^{[j]}} \dot{\zeta}^{[j]} \right) \cdot \frac{\partial \vec{r}_k}{\partial \zeta^{[i]}}. \qquad (3.138)$$

Consequently,

$$\frac{d}{dt} \frac{\partial L}{\partial \dot{\zeta}^{[i]}} = \sum_{k=1}^{N} m_k \left(\left(\sum_{j=1}^{M} \sum_{l=1}^{M} \frac{\partial^2 \vec{r}_k}{\partial \zeta^{[j]} \partial \zeta^{[l]}} \dot{\zeta}^{[j]} \dot{\zeta}^{[l]} + \sum_{j=1}^{M} \frac{\partial \vec{r}_k}{\partial \zeta^{[j]}} \ddot{\zeta}^{[j]} \right) \cdot \frac{\partial \vec{r}_k}{\partial \zeta^{[i]}} + \right.$$

$$\left. + \left(\sum_{j=1}^{M} \frac{\partial \vec{r}_k}{\partial \zeta^{[j]}} \dot{\zeta}^{[j]} \right) \cdot \left(\sum_{j=1}^{M} \frac{\partial^2 \vec{r}_k}{\partial \zeta^{[i]} \partial \zeta^{[j]}} \dot{\zeta}^{[j]} \right) \right). \qquad (3.139)$$

Besides,

$$\frac{\partial(\dot{\vec{r}}_k \cdot \dot{\vec{r}}_k)}{\partial \zeta^{[i]}} = 2\dot{\vec{r}}_k \cdot \frac{\partial \dot{\vec{r}}_k}{\partial \zeta^{[i]}} = 2 \left(\sum_{j=1}^{M} \frac{\partial \vec{r}_k}{\partial \zeta^{[j]}} \dot{\zeta}^{[j]} \right) \cdot \left(\sum_{j=1}^{M} \frac{\partial^2 \vec{r}_k}{\partial \zeta^{[j]} \partial \zeta^{[i]}} \dot{\zeta}^{[j]} \right), \qquad (3.140)$$

and

$$\frac{\partial \widetilde{U}}{\partial \zeta^{[i]}} = \sum_{k=1}^{N} \frac{\partial U}{\partial \vec{\mathbf{r}}_k} \cdot \frac{\partial \vec{\mathbf{r}}_k}{\partial \zeta^{[i]}}. \tag{3.141}$$

Substitution of (3.139)-(3.141) into (3.136) gives:

$$\sum_{k=1}^{N} \left(m_k \left(\sum_{j=1}^{M} \sum_{l=1}^{M} \frac{\partial^2 \vec{\mathbf{r}}_k}{\partial \zeta^{[j]} \partial \zeta^{[l]}} \dot{\zeta}^{[j]} \dot{\zeta}^{[l]} + \sum_{j=1}^{M} \frac{\partial \vec{\mathbf{r}}_k}{\partial \zeta^{[j]}} \ddot{\zeta}^{[j]} \right) + \frac{\partial U}{\partial \vec{\mathbf{r}}_k} \right) \cdot \frac{\partial \vec{\mathbf{r}}_k}{\partial \zeta^{[i]}} = 0. \tag{3.142}$$

Here $-\frac{\partial U}{\partial \vec{\mathbf{r}}_k}$ is the intramolecular force acting on atom A_k. Since the system, in fact, is not isolated, there is an additional force $\vec{\mathbf{g}}_k$ acting on A_k, if the atom is in contact* with surrounding media, such as water or cytosol. As before, we shall denote by $\vec{\mathbf{f}}_k := -\frac{\partial U}{\partial \vec{\mathbf{r}}_k} + \vec{\mathbf{g}}_k$ the sum of the intramolecular and hydration forces acting on atom A_k. Beside that, if atom A_k is exposed to solvent and moves, it is subjected to a drag force $\vec{\mathbf{d}}_k$.

For an isolated atom, the drag force predicted according to the Stokes' law is given by

$$\vec{\mathbf{d}}_k = -6\pi\mu R_k^{[w]} \dot{\vec{\mathbf{r}}}_k,$$

where μ is the dynamic viscosity of the solvent. Since any protein atom is at maximum only partially hydrated, we scale the drag force by the atom's hydration degree h_k and, additionally utilizing (3.137), obtain:

$$\vec{\mathbf{d}}_k = -6\pi\mu h_k R_k^{[w]} \sum_{j=1}^{M} \frac{\partial \vec{\mathbf{r}}_k}{\partial \zeta^{[j]}} \dot{\zeta}^{[j]}. \tag{3.143}$$

Supplementing (3.142) by forces, arising due to interaction with surrounding solvent, we obtain:

$$\sum_{k=1}^{N} \left(m_k \left(\sum_{j=1}^{M} \sum_{l=1}^{M} \frac{\partial^2 \vec{\mathbf{r}}_k}{\partial \zeta^{[j]} \partial \zeta^{[l]}} \dot{\zeta}^{[j]} \dot{\zeta}^{[l]} + \sum_{j=1}^{M} \frac{\partial \vec{\mathbf{r}}_k}{\partial \zeta^{[j]}} \ddot{\zeta}^{[j]} \right) - \right.$$
$$\left. -\vec{\mathbf{f}}_k + 6\pi\mu h_k R_k^{[w]} \sum_{j=1}^{M} \frac{\partial \vec{\mathbf{r}}_k}{\partial \zeta^{[j]}} \dot{\zeta}^{[j]} \right) \cdot \frac{\partial \vec{\mathbf{r}}_k}{\partial \zeta^{[i]}} = 0, \quad \forall\, i = \overline{1, M}. \tag{3.144}$$

To see how the values of different terms in (3.144) relate to each other, we shall recall the relation (3.128). Now, however, in general the both complementary branches $\mathcal{D}_{i_o i_\bullet}$ and $\mathcal{D}_{i_\bullet i_o}$ $((i_o, i_\bullet) = \iota(i))$ can move, rather than only the branch containing less atoms, as assumed in Proposition 3.51.

Let us assume that the carboxyl end of the polypeptide is still attached to the ribosome, and that all bonds complementary to those in the direction of flow from

---

the fixed bond obtained an infinite weight. Then the conditions of Proposition 3.51 are acceptable, and we shall assume for simplicity that they are fulfilled.

Now consider the second derivatives. Let $(i_\circ, i_\bullet) = \iota(i)$ and $(j_\circ, j_\bullet) = \iota(j)$. If $A_k \notin \mathcal{D}_{i_\circ i_\bullet} \cap \mathcal{D}_{j_\circ j_\bullet}$, then

$$\frac{\partial^2 \vec{r}_k}{\partial \zeta^i \partial \zeta^j} = 0. \tag{3.145}$$

Otherwise, either

$$\mathcal{D}_{j_\circ j_\bullet} \subseteq \mathcal{D}_{i_\circ i_\bullet} \tag{3.146}$$

or

$$\mathcal{D}_{i_\circ i_\bullet} \subset \mathcal{D}_{j_\circ j_\bullet} \tag{3.147}$$

holds.

If (3.146) is valid, then $A_{i_\circ}$ and $A_{i_\bullet}$ do not move with rotation of branch $\mathcal{D}_{j_\circ j_\bullet}$, therefore

$$\frac{\partial^2 \vec{r}_k}{\partial \zeta^i \partial \zeta^j} = \vec{e}_{i_\circ i_\bullet} \times \frac{\partial \vec{r}_{i_\bullet k}}{\partial \zeta^j} = \vec{e}_{i_\circ i_\bullet} \times \frac{\partial \vec{r}_k}{\partial \zeta^j} = \vec{e}_{i_\circ i_\bullet} \times (\vec{e}_{j_\circ j_\bullet} \times \vec{r}_{j_\bullet k}). \tag{3.148}$$

In the case of (3.147) we have:

$$\frac{\partial^2 \vec{r}_k}{\partial \zeta^i \partial \zeta^j} = \vec{e}_{j_\circ j_\bullet} \times (\vec{e}_{i_\circ i_\bullet} \times \vec{r}_{i_\bullet k}) \tag{3.149}$$

in accordance with equality of continuous mixed derivatives.

Let

$$\vec{e}_{i\circlearrowleft k} := \frac{\frac{\partial \vec{r}_k}{\partial \zeta^{[i]}}}{\left\| \frac{\partial \vec{r}_k}{\partial \zeta^{[i]}} \right\|}.$$

Then rearrangement and reduction of (3.144) yields:

$$\sum_{k=1}^{N} \left(m_k \left(\sum_{j=1}^{M} \sum_{l=1}^{M} \frac{\partial^2 \vec{r}_k}{\partial \zeta^{[j]} \partial \zeta^{[l]}} \dot{\zeta}^{[j]} \dot{\zeta}^{[l]} + \sum_{j=1}^{M} \frac{\partial \vec{r}_k}{\partial \zeta^{[j]}} \ddot{\zeta}^{[j]} \right) + 6\pi\mu h_k R_k^{[w]} \sum_{j=1}^{M} \frac{\partial \vec{r}_k}{\partial \zeta^{[j]}} \dot{\zeta}^{[j]} \right) \cdot \vec{e}_{i\circlearrowleft k} =$$

$$= \sum_{k=1}^{N} \vec{f}_k \cdot \vec{e}_{i\circlearrowleft k}, \quad \forall\, i = \overline{1, M}. \tag{3.150}$$

From (3.128), (3.145), (3.148), and (3.149) follows that

$$\left\| \frac{\partial \vec{r}_k}{\partial \zeta^{[i]}} \right\| \leq \|\vec{r}_{i_\bullet k}\| \leq D_p \quad \text{and} \quad \left\| \frac{\partial^2 \vec{r}_k}{\partial \zeta^i \partial \zeta^j} \right\| \leq \max(\|\vec{r}_{i_\bullet k}\|, \|\vec{r}_{j_\bullet k}\|) \leq D_p, \tag{3.151}$$

where D_p is the diameter of the protein molecule.

The dynamic viscosity of water at 298 K is 8.91×10^{-4} kg $\times$ (m $\times$ s)$^{-1}$ [14], which equals to 5.3659×10^{13} Da $\times$ (Å $\times$ s)$^{-1}$. In Figure 3.16 we see that atomic forces are typically of the order about 1 kcal$\times$(mol$\times$Å)$^{-1}$, which translates into 4.1869×10^{26} Da$\times$Å$\times$s^{-2}. Therefore, it is natural to expect $\dot{\zeta}^{[i]}$ and $\ddot{\zeta}^{[i]}$ of the orders around 10^{12} s^{-1} and 10^{24} s^{-2} respectively, if the terms in equation (3.150) have comparable magnitudes.

Besides, if the folding would follow a pathway close to the one obtained by the steepest descent in the dihedral angle space, it would imply that

$$\dot{\zeta}^{[i]} \propto \sum_{k\in\mathcal{N}_{i_o i_\bullet}} \vec{\mathbf{f}}_k \cdot \frac{\partial \vec{\mathbf{r}}_k}{\partial \zeta^{[i]}} = \sum_{k=1}^{N} \vec{\mathbf{f}}_k \cdot \vec{\mathbf{e}}_{i\circlearrowleft k} \left\| \frac{\partial \vec{\mathbf{r}}_k}{\partial \zeta^{[i]}} \right\|,$$

see Proposition 3.52. However, equation (3.150) suggests that any non-zero portion of the force projections to the directions of rotations (see the right-hand side), which is related to an angular velocity $\dot{\zeta}^{[j]}$, is, on the contrary, scaled by $1/\|\frac{\partial \vec{\mathbf{r}}_k}{\partial \zeta^{[j]}}\|$.

Therefore, a more handy scaling for the components of the energy gradient may help not only to avoid stiffness problems during energy minimization, but also to obtain a transformation path closer resembling a natural folding pathway.

CHAPTER 4

SiViProF Software

4.1 SiViProF – a new simulation software

On the basis of the model proposed in Chapter 3, a simulation and visualization cross-platform software SiViProF* is developed. It is written in C++ using OpenGL API for 3D visualization and QT library for graphical user interface. Figure 4.1 shows a screenshot of the program.

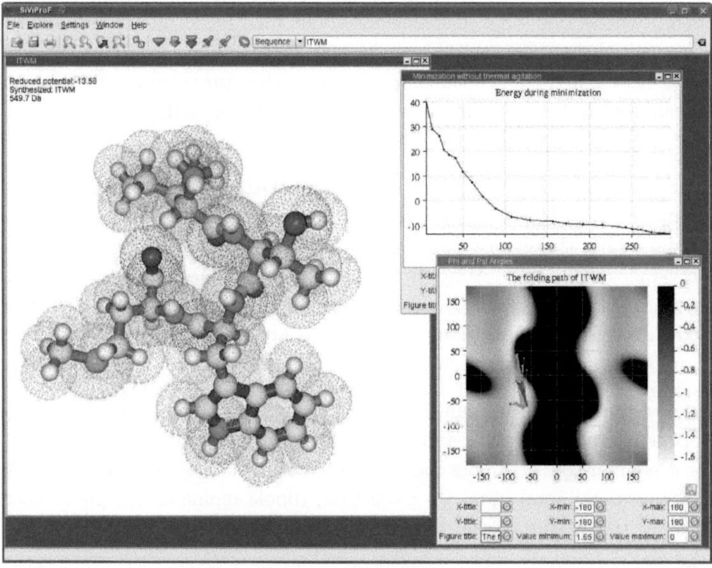

FIG. 4.1: A screenshot of SiViProF with the results of folding ITWM-tetrapeptide (without solvation) using the steepest descent without energy oscillations. The 1D plot shows how the energy changes during minimization (versus the maximal bond rotation in degrees). The highlighted subwindow depicts the folding path on a Ramachandran map.

---

*SiViProF stands for "SImulation and VIsualization of PROtein Folding".

Taking a sequence of amino acids coded by their conventional one-letter symbols (see Table A.2 in Appendix A) as an input, the program generates the corresponding polypeptide chain. Energy minimization is executed either upon appending each amino acid residue, or when the creation of the whole molecule is finished. The coordinates of the amino acids are generated immediately before or loaded from files with earlier computed structures. An attachment of a new residue is carried out so that a formed peptide group is disposed in the *trans* conformation (see Subsection 3.5.2). Visualization can be performed either as an animation of the minimization process, or only at the end, in order to save computational resources.

Initial atomic coordinates for amino acids are derived on the base of geometrical considerations motivated by known findings from quantum mechanics (see Subsections 1.5.1 and 1.5.2 for the theory, and Subsection 4.2.2 for a description of the algorithms for coordinate generation, suggested and used in this work). Special attention is paid to the correct chirality of amino acids. Structures containing pentagonal rings and variable dihedral angles are subsequently optimized. Such coordinate generation is considerably faster and provides better results than an optimization of a random coordinate set.

Energy minimization can be performed by different methods, some of which are described in Subsection 4.3.2. Conclusions about their performance and limitations are also given there. Computed protein structures can be saved in the format, specific for SIVIPROF, and loaded again for further processing. The list of the dihedral angles can be saved into a separate file and analyzed.

Experimentally determined structures from PDB (see Section 1.10) can be loaded by the program* (see Figure 4.2) for comparison with predicted structures and for refinement of the model parameters. PDB records often contain only coordinates of heavy atoms, since the signals from hydrogen atoms are too week to be detected by X-ray crystallography [87]. Apart from that, there is some discrepancy in naming hydrogen atoms, which generally prohibits loading related coordinates, if they are provided. Therefore, the positions of hydrogens are reconstructed by the program according to the Algorithm 4.4 described in Subsection 4.2.2.

Molecules of other chemicals can be also generated by the program, provided that their formulas are given in the appropriate format, described in Section A.3. This enables a comparison of dipole moments or verification of solvation energies on the base of experimental data. In particular, dipole moments can be utilized for evaluation of the quality of the computed partial charges used in simulations.

The following molecular visualization modes are implemented: space-filling models with representation of atoms as van der Waals spheres (Fig. 4.2(a)), also including atomic solvation grids (Fig. 4.3(b)), ball and stick models with reproduction of

---

*There is still a number of limitations that can be eliminated in future. Thus, the structures containing posttranslationally modified residues or multiple polypeptide chains can not be loaded completely. Associated water molecules and ions are ignored.

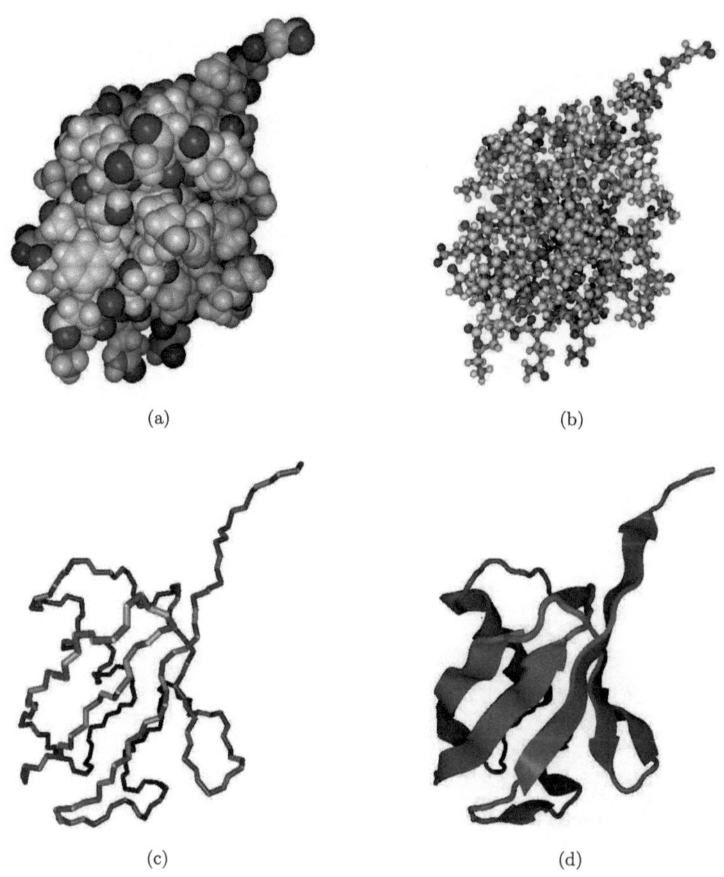

FIG. 4.2: The native structure of ubiquitin, visualized by SiViProF. An interpretation of colors is given in Table A.1. Atomic coordinates are obtained from RCSB Protein Data Bank, record 1UBQ by S. Vijay-Kumar *et al.* [123]. The positions of hydrogen atoms are reconstructed by SiViProF. (a) The space-filling model. (b) The all-atom ball-and-stick model. (c) The ball-and-stick model showing only the configuration of the main chain. (d) The ribbon model depicting the secondary structure (see Subsection 1.6.2 for details).

atoms and bonds in a form of spheres and cylinders (Fig. 4.2(b)), also with dotted van der Waals spheres (Fig. 4.3(a)), stick and ribbon models, depicting only the main chain conformation in one or several colors (Fig. 4.2(c, d)). There is also an option to show all main chain atoms with bonds and to hide side chains, in order to facilitate an exploration of the secondary structure. Besides, there are different modes depicting atomic and twisting forces (Fig. 4.3(c)), interatomic interaction energies (Fig. 4.3(d)), atomic partial charges (Fig. 4.3(e)), or hydration (Fig. 4.3(e)).

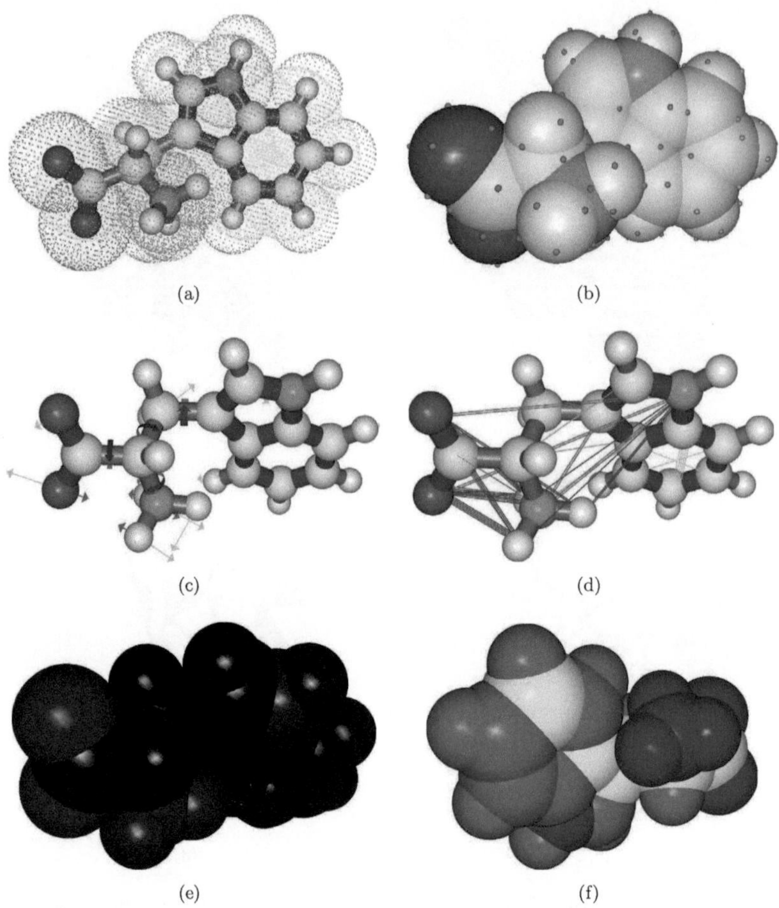

(a)   (b)

(c)   (d)

(e)   (f)

FIG. 4.3: A tryptophan molecule constructed and visualized by SIVIPROF. (a) The ball-and-stick model, where the atomic van der Waals spheres are outlined by dots. (b) The space-filling model, additionally showing atomic solvation grids. (c) The ball-and-stick model with depicted electrostatic and van der Waals forces (straight red and yellow arrows respectively), as well as twisting forces (bended arrows around bonds). (d) The ball-and-stick model showing electrostatic and van der Waals interactions between atoms. Each line corresponds to 0.1 kcal/mol. Red and blue lines correspond to the repulsive and attractive electrostatic interactions respectively. Yellow and green lines - to the repulsive and attractive van der Waals interactions. (e) The space-filling model depicting atomic partial charges. The explanation for colors is given on p. 31. (f) The space-filling model showing favorable and unfavorable hydration approximated as described in Section 3.4. Red color denotes hydrated hydrophobic atoms, yellow – non-hydrated hydrophobic atoms, blue – hydrated hydrophilic atoms, and orange is reserved for non-hydrated hydrophilic atoms.

The radii of atom spheres are either equal to van der Waals radii, used in computations, or amount to the one third of those, if the bonds are also depicted. Colors and bond radii, as well as radii of solvation grid points, can be prescribed by the user. Additionally, one can activate the fog or change the opacity of objects. Rotation and zooming the three-dimensional molecule image is possible.

In the modes with visualization of forces or interactions, arrows and lines are always depicted in a way that allows them to be better visible to the viewer. This is done by taking into account the position of the camera. There are three different options for visualization of interactions: the first two imply that the quantitative information about the interaction intensity is given by the number of lines (at maximum ten). Each line corresponds to either 0.1 kcal/mole (see Figure 4.3(d)) or $\bar{R}T$, depending on the choice. The last option permits the output of the numerical information about each interaction.

Beside the simulation and visualization functions, some procedures for exploration of protein free energy landscapes and analysis of the contributions of different forces were developed. One can deactivate certain atoms, change the geometry of the molecule and observe the resulting energy changes. Energy profiles and surfaces can be plotted for selected dihedral angles (see Figures 1.25, 1.27-1.30, and 4.7-4.8).

Apart from that, SiViProF includes procedures for visualization of standard atomic orbitals (see Figures in Subsection 1.5.1) and large molecular complexes, which can be cut and and shown in slices (see Figures in Subsection 1.8.1).

4.2 Implementation of the model

In this section we shall discuss the implementation of the model described in the previous chapter. In particular, the algorithms needed for generation of initial coordinates, as well as some other algorithms necessary but not directly related to energy minimization, which were developed and implemented in course of this work, shall be discussed. The algorithms, introduced here and in the following sections, are simplified versions of the implemented functions. Technical details, such as allocation and release of memory or declaration and initialization of variables, are omitted. Minimization of energy is a subject of a separate discussion elucidated in Subsection 4.3.2 together with simulation results.

4.2.1 Listing degrees of freedom

For some of the implemented minimization algorithms it is necessary to acquire a list of bonds that can be rotated together with originated branches (see definitions given in Section 3.3). Since the same change in a dihedral angle can be achieved by rotating any of two complementary branches, only the one which contains less atoms should be included in the list. There is no sense to rotate a branch consisting of only one atom or a branch complementary to it. Double or partially double bonds, as well as bonds inside a ring, should not be considered either, since the corresponding dihedral angles are fixed.

Algorithm 4.1 assigns to each bond b_{ij} a weight w_{ij} (according to the definition given by equation (3.84)), which expresses the number of atoms that must be moved, if the branch originated by b_{ij} is dislocated*. Bonds that belong to a ring obtain zero weight. Initially the weight -1 is assigned to each bond, except for the bonds to hydrogen atoms or to the *dummy atom*, which represents the temporary carboxyl end of the protein during the synthesis. The latter bonds get a weight equal to one. The weight of the bond is considered to be *permanent*, if it does not change upon appending a new amino acid.

The algorithm is performed recursively for each bond b_{ij} in the direction of flow from the first considered bond or for the bond complementary to it. The latter may happen, if b_{ij} belongs to a ring. The considered bond b_{ij} is given by atoms A_i and A_j, introduced as input parameters. The procedure is first called from Algorithm 4.2, which preliminary sets the branch origin A_o and some other external variables. N denotes the current number of atoms in the molecule, and N_m is the instant number of atoms that have to be moved when the branch originated by the considered bond is displaced. N_l and N_b account for the numbers of loop closures and of atoms in out-of-loop branches.

ALGORITHM 4.1 (DETERMINATION OF BOND WEIGHTS)

For given atoms A_i and A_j do the following:

   1) if $A_j = A_o$, increase N_l by one, set $N_b = 0$, and go to the Step 8,

   2) note the current N_m value: $N_i = N_m$,

   3) increase N_m by one,

   4) if $N_l > 0$ increase N_b by one,

   5) mark atom A_j,

   6) for each atom A_k $(k \neq i)$ that is bonded to A_j do the following:
     if A_k is marked, block the bond b_{kj}, set $N_b = w_{jk} = w_{kj} = 0$,
                 and increase N_l by one,
    otherwise, if the bond b_{jk} is blocked, unblock it and decrease N_l by one,
       otherwise, if $w_{jk} < 1$, call the whole procedure
             with A_j <u>as</u> A_i and A_k <u>as</u> A_j,
      otherwise, increase N_m by w_{jk},

   7) erase the mark from atom A_j,

---

*Since dislocation is not restricted to branch rotation, the atom A_j is also counted in the weight of bond b_{ij}.

8) if $N_l > 0$, choose one of the following options:
$$\text{if } N_b = 0, \text{ set } w_{ij} = w_{ji} = 0,$$
otherwise decrease N_b by one,

9) if $w_{ij} \neq 0$, set $w_{ij} = N_m - N_i$, $w_{ji} = N - w_{ij}$ and raise permanent weight flag for the bond b_{ij}.

Step 1 in Algorithm 4.1 is necessary for cases, when A_o belong to a ring. If the numeration of atoms assures that the bond weight search does not start inside a ring, this step can be omitted.

Algorithm 4.2 uses the bond weights, assigned by Algorithm 4.1, to build list $\mathcal{B}$ of the bonds that are relevant for energy minimization in dihedral angle space. The determination of the bond weights together with the bond list formation should be performed before generation of initial atomic coordinates, since the zero bond weight is utilized for identification of ring atoms in the localization procedure (see Section 4.2.2). The algorithm is called with a single parameter N_p representing the number of atoms during the previous bond list formation, if it was already performed. N_p is set to zero if the bond list is to be generated for the first time.

ALGORITHM 4.2 (BUILDING THE LIST OF RELEVANT BONDS)

1) note the current number of bonds in list $\mathcal{B}$ for use in other algorithms,

2) clear list $\mathcal{B}$,

3) if $N_p > 0$, for each atom A_i that is not hydrogen and such that
$$i < N_p, \tag{4.1}$$
find all the bonds w_{ij} with permanent weight and set $w_{ji} = N - w_{ij}$,

4) set $N_m = N_l = N_b = 0$, let A_o be the last atom and perform Algorithm 4.1 starting with A_o <u>as</u> A_i and an atom, bonded to it, <u>as</u> A_j, in order to determine all bond weights,*

5) for each atom A_i that is not a hydrogen or the dummy atom, consider each bonded atom A_j, such that
$$w_{ij} > 1, \ w_{ji} > 1, \tag{4.2}$$
$$j < i, \tag{4.3}$$
and excluding $j \simeq i$,

determine the bond status (ϕ, ψ or other) basing on the roles[†] of the bonded atoms, and add the bond
$$b_{ij}, \quad \text{if } w_{ij} < w_{ji},$$
$$b_{ji}, \quad \text{otherwise,}$$

---

*In case of an incomplete molecule it has to be the dummy atom.

†A dihedral angle ϕ is related to a bond between a nitrogen of a peptide group and an α-carbon, and a dihedral angle ψ corresponds to a bond connecting C_α and C', see Section 1.6.1.

together with its status and its residue number to list $\mathcal{B}$.

If the minimization is performed before the polypeptide is completely synthesized, the list of relevant bonds has to be rebuilt, and the weights that are changed have to be reassigned. This is done in Step 3 of Algorithm 4.2. Since for any two connected atoms A_i and A_j the equality

$$w_{ij} + w_{ji} = N$$

holds, permanent weights can be used for reconstruction of other weights in the earlier synthesized part of the molecule. Atoms are numbered consequently, starting from zero. Therefore, Condition 4.1 eliminates excessive consideration of the atoms from the new part.

The determination of the bond weights for the newly created part is performed in Step 4. In the current implementation, the dummy atom is always the last one, and it temporary has only one bond. Therefore the bonds in direction of flow from the bond of the dummy atom get permanent weight, when Algorithm 4.1 is called starting with the last atom as a branch origin. Herewith, the already determined weights are used for computation of the remaining ones without repeated examination of the explored part of the molecule.

Conditions (4.2) in Step 5 exclude registration of the bonds that belong to a ring or originate a branch consisting of only one atom. Condition (4.3) assures that the bond is not included to the list two times. Step 5 of Algorithm 4.2 can be performed as a part of Step 9 in Algorithm 4.1, but this would result in an inconvenient sequence of bonds in $\mathcal{B}$.

4.2.2　Generation of initial coordinates

As already mentioned in Section 4.1, PDB records may not provide information about locations of hydrogen atoms. Algorithm 4.4 can be used for reconstruction of hydrogen atom positions, when the coordinates of other atoms are known. It is also used in combination with other algorithms for overall coordinate generation.

Definition 4.3

In the following, we shall refer to an atom as *located*, if its position is already specified. Otherwise we shall call it *unlocated*. By saying that we would like to *locate* an atom, we shall imply that we are going to specify its position with fulfillment of the following criteria:

a) the bonds to other located atoms obtain the corresponding equilibrium bond lengths,

b) the bond angles that are determined by this process acquire their equilibrium values,

c) atom groups that must remain in one plane are located respectively.

Besides, we shall take care of chirality where necessary. As discussed in Subsection 1.3, there are not many chiral centers in amino acids. The correct chirality can be set immediately on Step 3.3 of Algorithm 4.4 or later only for selected chiral centers, as described in Subsection 4.2.3.

Algorithm 4.4 (Location of hydrogen atoms)

For a designated unlocated atom A_i:

1) *find a located atom A_j, connected to A_i,*

2) *determine the sets $\mathcal{L}$ and $\mathcal{U}$ of the numbers of located and unlocated atoms bound to A_j,*

3) *depending on the cardinalities of $\mathcal{L}$ and $\mathcal{U}$, choose one of the following options*[*]:*

   3.1) card $\mathcal{L} = 1$:
   if $j \simeq k$, $k \in \mathcal{L}$, then try to find any located atom A_m $(m \neq j)$ bonded to A_k and, in case of success, set

   $$\vec{\mathbf{r}}_i = \vec{\mathbf{r}}_j + \frac{l_{ij}}{\|\vec{\mathbf{r}}_{kj}\|}\vec{\mathbf{r}}_{kj}\llcorner^{\vec{\mathbf{r}}_{km}}(\alpha_j - \pi),$$

   otherwise, set

   $$\vec{\mathbf{r}}_i = \vec{\mathbf{r}}_j + \frac{l_{ij}}{\|\vec{\mathbf{r}}_{kj}\|}\vec{\mathbf{r}}_{kj}\llcorner(\pi - \alpha_j), \ k \in \mathcal{L}, \tag{4.4}$$

   3.2) card $\mathcal{U} = 1 \wedge$ card $\mathcal{L} \neq 1$: set

   $$\vec{\mathbf{r}}_i = \vec{\mathbf{r}}_j + l_{ij} \sum_{k \in \mathcal{L}} \frac{\vec{\mathbf{r}}_{kj}}{\|\vec{\mathbf{r}}_{kj}\|},$$

   3.3) card $\mathcal{L} =$ card $\mathcal{U} = 2$ (without loss of generality, let $l, k \in \mathcal{L}$ and $i, m \in \mathcal{U}$): set

   $$\vec{\mathbf{r}}_i = \vec{\mathbf{r}}_j + l_{ij}\mathbf{T}^{\vec{\mathbf{r}}_{jk}}_{\vec{\mathbf{r}}_{jl}}\vec{\mathbf{v}}_l$$

   and

   $$\vec{\mathbf{r}}_m = \vec{\mathbf{r}}_j + l_{jm}\mathbf{T}^{\vec{\mathbf{r}}_{jk}}_{\vec{\mathbf{r}}_{jl}}\vec{\mathbf{v}}_r,$$

   if the atoms A_i, A_l, and A_m are to appear in counterclockwise order when viewed from A_k in the direction of A_j. Otherwise the roles of i and m have to be exchanged. $\mathbf{T}^{\vec{\mathbf{r}}_{jk}}_{\vec{\mathbf{r}}_{jl}}$, $\vec{\mathbf{v}}_l$, and $\vec{\mathbf{v}}_r$ are defined by equations (3.116) and (3.117).

---

[*]If only the locations of the hydrogen atoms in a protein are to be determined, A_j is necessarily bound to at least one located atom. In case of general coordinate generation the fulfillment of the latter condition has to be ensured by the atom selection sequence, see Algorithm 4.5. Since the elements that constitute proteins form at maximum four bonds, the proposed choice is comprehensive.

Thus, Algorithm 4.4 assumes that equilibrium bond angles correspond to those expected for standard hybridization states (see Subsection 1.5.1). Therefore, if the directions of two or three bonds of an atom are known, the directions of the remaining bonds are deduced automatically, with no request for the equilibrium bond angle parameter α_j. Since sp^3-hybridized nitrogen atoms are usually protonated in solution (see Section 1.4), they can be naturally treated as having four covalent bonds due to delocalization of electrons.

The following algorithm is used in SIVIPROF as a framework for generation of atomic coordinates for amino acids and some other small molecules. It makes use of the below described Algorithms 4.7-4.6 to locate ring atoms and calls the above described Algorithm 4.4 to determine the positions of other atoms.

ALGORITHM 4.5 (OVERALL COORDINATE GENERATION)

1) *Set the first atom A_0 into the coordinate origin,*

2) *choose another atom A_b, bonded to A_0, and set $\vec{r}_b = (l_{01}, 0, 0)^T$,*

3) *perform the following steps consequently for all atoms as long as there are unlocated atoms left:*

   3.1) *if the considered atom A_i is unlocated, try to locate it according to the Algorithm 4.4; in case of failure on Step 1 of the Algorithm 4.4 proceed to the next atom,*

   3.2) *for any atom A_j bound to A_i: if $w_{ij} \neq 0$, and A_j is unlocated, locate it using Algorithm 4.4,*

   3.3) *if A_i belongs to a ring, locate ring atoms according to the Algorithm 4.6 starting with A_i.*

The next Algorithm is used in SIVIPROF for generation of coordinates for ring atoms. It assumes that the bond weights are already assigned by Algorithm 4.1 and that the atom numbering does not start inside the ring. That is, if Algorithm 4.6 is called from Algorithm 4.5 for a specified atom A_r, the set of located atoms connected to A_r is not empty. Otherwise the direction of the ring center could be chosen randomly or parallel to one of the coordinate axes.

ALGORITHM 4.6 (LOCATION OF RING ATOMS)

For a specified located ring atom A_r find a connected atom A_u that is unlocated and belongs to a ring (i.e., $w_{ru} = 0$), and perform the following steps:

1) *set $w_u = 1$,*

2) *perform Algorithm 4.7 starting with A_r <u>as</u> A_i and A_u <u>as</u> A_j, to find a single ring and to determine the loop weights of ring atoms,*

3) *establish the set $\mathcal{O} = \{o_1, \ldots, o_{N_r}\}$ of the ring atom numbers as following: set $o_{N_r} = r$, and subsequently decreasing i from $N_r - 1$ to 1 choose the atom A_{o_i} connected to $A_{o_{i+1}}$, so that $w_{o_i} = i$,*

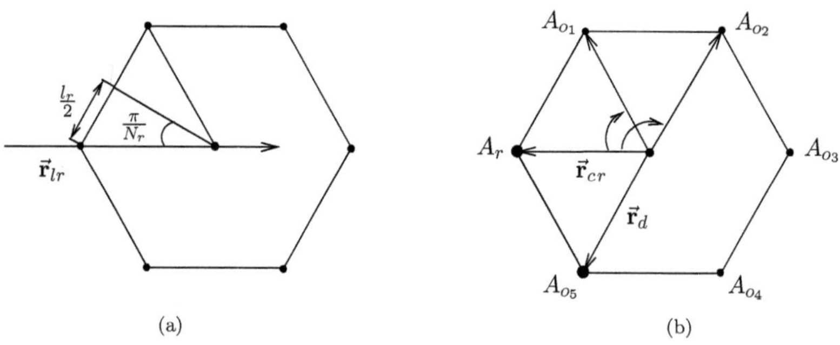

FIG. 4.4: Location of ring atoms (see Algorithm 4.6): (a) determination of the center position for card $\mathcal{L} = 1$, (b) computation of positions for unlocated ring atoms in case when A_{o5} is located (as well as A_r).

4) *find the location of the loop center as*

$$\vec{\mathbf{r}}_c = \vec{\mathbf{r}}_r + \frac{l_r \sum_{l \in \mathcal{L}} \vec{\mathbf{r}}_{lr}}{2\|\sum_{l \in \mathcal{L}} \vec{\mathbf{r}}_{lr}\| \sin \frac{\pi}{N_r}},$$

where l_r is the average bond length in the ring, and $\mathcal{L}$ is the set of numbers of located atoms connected to A_r, that do not belong to this ring (see Figure 4.4 (a)),

5) *define the vector $\vec{\mathbf{r}}_d$ giving the direction for deflection (see Figure 4.4 (b)):*

$$\vec{\mathbf{r}}_d = \begin{cases} \operatorname{sign}(\frac{N_r}{2} - k)\vec{\mathbf{r}}_{co_k}, & \text{if exists a located atom } A_{o_k},\ o_k \in \mathcal{O} \setminus \{o_{N_r}\}, \\ & \text{such that } \vec{\mathbf{r}}_{cr} \nparallel \vec{\mathbf{r}}_{co_k}, \\ \vec{\mathbf{r}}_{cr} \llcorner \left(\frac{2\pi}{N_r}\right), & \text{otherwise,} \end{cases}$$

6) *determine the position of each unlocated atom A_{o_n}, $o_n \in \mathcal{O}$, as*

$$\vec{\mathbf{r}}_{o_n} = \vec{\mathbf{r}}_c + \vec{\mathbf{r}}_{cr} \llcorner^{\vec{\mathbf{r}}_d} \left(\frac{2\pi n}{N_r}\right).$$

The first option in Step 5 is necessary, for example, in cases when the other part of a double ring is already located. The structure of a proline ring has to be subsequently optimized or generated by another method, if specific puckering is desired.

The idea of Algorithm 4.7 is to find the shortest loop in rings, starting from a given atom A_r in the direction of the second specified atom. The algorithm represents a recursive function, which is called with three parameters. Two of them refer respectively to the previously and to the currently considered atom A_i and A_j, while the last one – to the loop origin A_r. In course of the function execution, a loop weight w_j, equal to the number of atoms between A_r and A_j (inclusive A_j), is assigned to atom A_j. Respectively, w_r is set equal to zero.

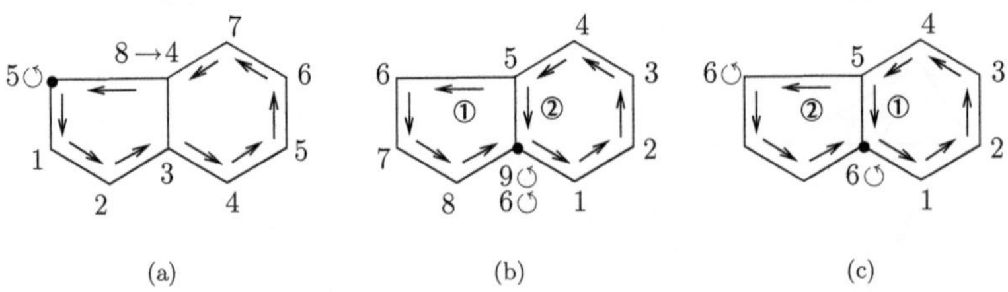

FIG. 4.5: Determination of the shortest loop. • marks the loop origin, simple numbers designate the loop weights of the atoms, and numbers in circles show the order of bond selection for passage. ↻ denotes the call return. (a) Identification of the shortcut. (b) Reassignment of the number of ring atoms. (c) No reassignment is performed if the shortest way was found first.

ALGORITHM 4.7 (LOOP EXPLORATION)

For specified atoms A_i and A_j do the following:

1) *if $A_j = A_r$, set the final number of atoms in the ring $N_r = w_j$ and*
   return to the call place,
   otherwise proceed to the next steps,

2) *mark atom A_j,*

3) *if one of the connected atoms A_k is marked and has the loop weight $w_k < w_j - 1$, set $w_j = w_k + 1$,*

4) *for any connected atom A_m ($m \neq i$) which is not marked and belongs to a ring, i.e. $w_{jm} = 0$, set $w_m = w_j + 1$ and call the steps 1-5 for $A_i = A_j$ and $A_j = A_m$,*

5) *erase the mark from the atom A_j.*

Marking in Step 2 prevents returning to the passed atoms, and erasing the marks is necessary for further processing. Step 5 is performed when all calls from Step 4 are returned. Step 3 is necessary for double rings to detect the shortest loop. If after branching the longer way was chosen, the shortcut can be identified at this step (Fig. 4.5(a)), unless the loop origin A_r is encountered before this check was performed. The latter can happen when A_r is shared between two rings (Fig. 4.5(b)). The origin is not marked, since it would prevent its consideration on Step 4, therefore the shortcut check is not performed for it. Nevertheless the shortcut is recognized when the fulfillment of Step 4 is continued at the branching point, since the rings have only one shared bond. In this case the condition of Step 1 is fulfilled second time and the correct value is assigned to N_r. Only after that the call is returned. The return prevents the reassignment of the value of N_r if the shortest way was found first (Fig. 4.5(c)).

4.2.3 Chirality correction

The chirality can be examined as follows. Assume that atom A_j is the considered chiral center, and atoms A_i, A_k and A_m must appear in the clockwise order when viewed from atom A_l to A_j ($i \bowtie j \bowtie k$, $m \bowtie j \bowtie l$). Then A_i and A_m must be at different sides of the plane passing through A_k, A_j and A_l, and, moreover, must hold:

$$(\vec{\mathbf{r}}_{jk} \times \vec{\mathbf{r}}_{jl}) \cdot \vec{\mathbf{r}}_{jm} > 0 \quad \text{and} \quad (\vec{\mathbf{r}}_{jk} \times \vec{\mathbf{r}}_{jl}) \cdot \vec{\mathbf{r}}_{ji} < 0.$$

It is sufficient to check only one of these conditions, if the bond directions at least approximately resemble those expected for an sp^3-like configuration.

Algorithm 4.8 describes an implementation of a chirality correction on the example of an α-carbon chiral center[*]. It can be used also for other chiral centers when atom identifiers are replaced appropriately. It assumes that the bond angles have to be corrected also, otherwise a simple reflection may be sufficient, if there is no other chiral center in the molecule.

The idea of the algorithm is, at first, to move the shortest of the two main chain branches in order to fit the bond angle $N - C_\alpha - C'$, and then to reconstruct the positions of the hydrogen and of the β-carbon. The latter is used for an appropriate dislocation of the whole side chain brach. For branch dislocations Algorithm 4.10 is used. In the following, $\angle C_\alpha$ denotes the reference bond angle for the α-carbon.

Algorithm 4.8 (Chirality correction)

For a specified C_α *find the connected* C', N, H *and* C_β *and do the following:*

1) *if* $w_{C_\alpha C'} > w_{C_\alpha N}$, *determine the new direction* $\vec{\mathbf{d}} = \vec{\mathbf{r}}_{C_\alpha N} \overset{\angle C_\alpha}{\vee} \vec{\mathbf{r}}_{C_\alpha C'}$
   and move branch $\mathcal{D}_{C_\alpha N}$ *using rotation matrix* $\mathbf{M}_{\vec{\mathbf{d}}}^{\vec{\mathbf{r}}_{C_\alpha N}}$,

   otherwise, set $\vec{\mathbf{d}} = \vec{\mathbf{r}}_{C_\alpha C'} \overset{\angle C_\alpha}{\vee} \vec{\mathbf{r}}_{C_\alpha N}$
   and move branch $\mathcal{D}_{C_\alpha C'}$ *using rotation matrix* $\mathbf{M}_{\vec{\mathbf{d}}}^{\vec{\mathbf{r}}_{C_\alpha C'}}$,

2) *set* $\vec{\mathbf{r}}_H = \vec{\mathbf{r}}_{C_\alpha} + l_{C_\alpha H} \mathbf{T}_{\vec{\mathbf{r}}_{C_\alpha N}}^{\vec{\mathbf{r}}_{C_\alpha C'}} \vec{\mathbf{v}}_r$,

3) *set* $\vec{\mathbf{d}} = \mathbf{T}_{\vec{\mathbf{r}}_{C_\alpha N}}^{\vec{\mathbf{r}}_{C_\alpha C'}} \vec{\mathbf{v}}_l$,

4) *move branch* $\mathcal{D}_{C_\alpha C_\beta}$ *using rotation matrix* $\mathbf{M}_{\vec{\mathbf{d}}}^{\vec{\mathbf{r}}_{C_\alpha C_\beta}}$.

Lemma 4.9

Let A_o *be the branch origin and* $\mathbf{M}$ *be a matrix of rotation relative to* A_o. *Then the new position* $\vec{\mathbf{r}}_i'$ *of a branch atom* A_i *is given by*

$$\vec{\mathbf{r}}_i' = (\mathbf{E} - \mathbf{M})\vec{\mathbf{r}}_o + \mathbf{M}\vec{\mathbf{r}}_i,$$

where $\mathbf{E} \in \mathbb{R}^{3 \times 3}$ *denotes a unit matrix.*

---

[*]See Section 1.3 for a discussion of chirality in amino acids.

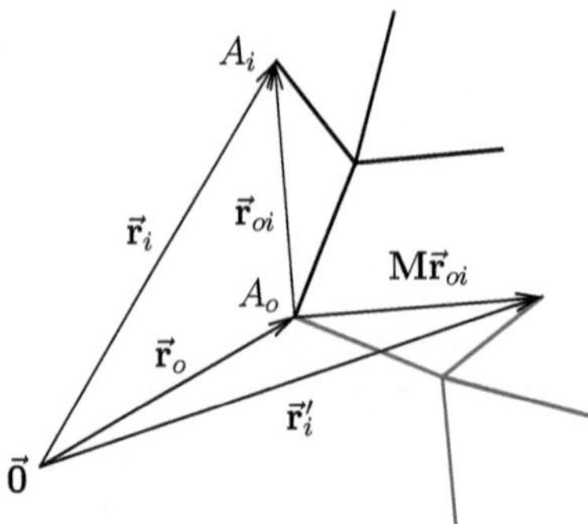

FIG. 4.6: Branch dislocation.

Proof:

$$\vec{\mathbf{r}}_i' = \vec{\mathbf{r}}_o + \mathbf{M}\vec{\mathbf{r}}_{oi} = \vec{\mathbf{r}}_o + \mathbf{M}(\vec{\mathbf{r}}_i - \vec{\mathbf{r}}_o) = (\mathbf{E} - \mathbf{M})\vec{\mathbf{r}}_o + \mathbf{M}\vec{\mathbf{r}}_i,$$

see Figure 4.6. □

The purpose of Algorithm 4.10 is to rotate branch $\mathcal{D}_{od}$ for given atoms A_o and A_d. The algorithm uses atomic transition numbers n_i, $n \in \mathcal{N}_{od}$ to mark atoms that have been already moved. This algorithm can be used both for chirality corrections and for branch rotations during energy minimization.

ALGORITHM 4.10 (BRANCH DISLOCATION)

For specified atoms A_o, A_d and rotation matrix $\mathbf{M}$ do the following:

1) *determine the translation vector* $\vec{\mathbf{v}} = \vec{\mathbf{r}}_o - \mathbf{M}\vec{\mathbf{r}}_o$,

2) *increase the global transition number N_t by one,*

3) *set $n_o = N_t$,*

4) *perform recursively the steps 4.1-4.2 starting with A_d as A_i:*

   4.1) *set $n_i = N_t$*

   4.2) *for each atom A_j bonded to A_i, such that $n_j \neq N_t$, determine the new position $\vec{\mathbf{r}}_j'$ as*

   $$\vec{\mathbf{r}}_j' = \vec{\mathbf{v}} + \mathbf{M}\vec{\mathbf{r}}_j \tag{4.5}$$

   and call the procedure 4.1-4.2 for A_j as A_i

5) *if A_d should be moved*, determine its new coordinates using (4.5).*

---

*If the branch is rotated about vector $\vec{\mathbf{r}}_{od}$, there is no need to recompute the coordinates of A_d, since they do not change.

4.2.4 Overall organization

Two force fields were implemented in SiViProf. The one given by equation (3.87) with the associated definitions, additionally supplemented by the terms modeling ribosomal restriction (specified in (3.109)) and distortions related to disulfide bridges (described by (3.124) and (3.125)), together with the related forces, and another one of the type (2.2), supplemented by the same terms, for energy minimization in the space of Cartesian coordinates.

The latter force field differs from the classical form (2.2) in a number of details, apart from the mentioned additional terms. In particular, the electrostatic potential involves screening functions, as in (3.87). The difference in the angle bending term is that the constants $k_i^{[a]}$ and α_i mostly depend only on the central atom A_i. This minor simplification is justified by quantum mechanical speculations given in Section 1.5 and through comparison of parameters used in other force fields[*]. The torsional term takes into consideration only double and partially bonds that do not belong to any ring. This term is aimed to prohibit twisting about such bonds. Twisting about single bonds is possible, and the desired shape of the associated energy profile is achieved due to the repulsive contribution of the van der Waals interactions of the neighboring bonded atoms. Out-of-plane bending is hindered through the utilization of appropriate bond-angle reference values for atoms in five-member aromatic rings, therefore no extra terms for potential due to out-of-plane bending are necessary.

In the both implemented force fields, the particular form of the Lennard-Jones potential together with the associated parameters is adopted from [94]. A cutoff of 4 Å is used for evaluation of the van der Waals interactions. This condition assumes that only the atoms that are immediately in contact are involved into interaction. Besides the reduction of computational costs, it facilitates an estimation of the van der Waals interaction with surrounding solvent without explicit consideration of its molecules, using available information about solvent-exposed surface area from the hydration model, proposed in this work (see Section 3.4). Computations show that the error introduced by this cutoff is sufficiently small.

In the current implementation, a disulfide bridge is formed when two sulfur atoms approach each other to a distance equal or smaller than the sum of their van der Waals radii, and breaks when the energy of the related geometry distortions (bond stretching and angle bending) exceeds the mean enthalpy of disulfide bonds (64.05 kcal/mol [14]). Although this is a simplified picture, this realization is helpful for a more appropriate reproduction of protein structures.

Hydrogen bonds are implemented similar to the Tripos 5.2 force field [94] (see Subsection 2.2.4). Thus, no van der Waals interaction is computed between the hydrogens in donor groups and potential hydrogen bond acceptors.

---

[*]The values of the utilized parameters are listed in Appendix D.

As suggested in Subsection 3.5.2, a collection of the amino acids that are necessary for synthesis is created preliminary. If the protein structure is to be loaded from a file, the amino acids are produced without specification of atomic positions. Otherwise the coordinates are generated according to the description given in Subsection 4.2.2, or loaded from files with precomputed amino acid structures. These coordinates are used for calculation of the new atom positions after inclusion of the corresponding residue into the protein.

Before a non-leading amino acidis appended, its amino group is converted into the non-ionized form. One of the oxygen atoms at the carboxyl end of the growing chain is deleted, and its bond to C' is replaced by a peptide bond, connecting the carbon with the nitrogen of the amino group of the attaching residue. The coordinate transformation for the new residue is performed as described in Subsection 3.5.2.

When the assignment (4.4) is performed in Algorithm 4.4, the obtained bond direction has a random contribution and, in general, requires optimization, when the whole structure is generated. In particular, if Algorithm 4.4 is used for determination of positions for only the hydrogen atoms (in the case when other coordinates are loaded from PDB), the bond b_{kj} between the two previously located atoms (i.e. not the bond b_{ji} to the unlocated atom, cf. (4.4)) can be added to the list of bonds relevant for energy minimization. Thus, the hydrogen positions that are not uniquely defined can be efficiently optimized, while other atoms remain fixed at the specified positions.

In order to download the coordinates of non-hydrogenic atoms from PDB, the appropriate atom labels must be assigned. They consist of a two-character right-justified atom name followed by an alphabetic symbol and the branch number. The alphabetic symbol is essentially the same as the conventional identifier of the given atom (see Table A.2 in Appendix A), except that the Greek letter is replaced by the corresponding capital Latin letter. While the naming for the common main-chain part is relatively simple to implement (N, C_α, C', and O obtain respectively the labels " N "," CA "," C ", and " O "), labeling the remaining atoms requires an additional discussion.

Algorithm 4.11 is used in SIVIPROF for generation of PDB atom labels for all amino acids with exception of glycine.

ALGORITHM 4.11 (ASSIGNMENT OF PDB ATOM LABELS)

    1) add the number of the β-carbon to an empty list $\mathcal{O}$ of atom numbers,*

    2) set the label of the β-carbon equal to " CB " and the alphabetic index to "G",

    3) repeat the steps 3.1-3.5 while the current list $\mathcal{O}$ is not empty,

---

*According to the numeration used in the current implementation, β-carbon has number 9 for all amino acids.

> *3.1) read the current branch number from the last position in the label of atom A_{o_1}, where o_1 is the first number in list $\mathcal{O}$ (space assumes the branch number one),*
>
> *3.2) for each atom A_{o_k}, $o_k \in \mathcal{O}$, consider consequently all directly connected atoms that are not yet named and not hydrogens, and perform the steps 3.2.1-3.2.3 for them:*
>
>> *3.2.1) set the branch number equal to the maximum between the current value and the branch number of A_{o_k},*
>>
>> *3.2.2) name the atom by the string constituted consequently from the following symbols: a space character*, the atom type, the alphabetic index and the branch number,*
>>
>> *3.2.3) increase the branch number counter,*
>
> *3.3) increase the alphabetic index as appropriate,*
>
> *3.4) if there is only one atom in list $\mathcal{O}$, and its branch number is one, rewrite the corresponding position in the label by a space character,*
>
> *3.5) assign list $\mathcal{N}$ to list $\mathcal{O}$.*

For a few amino acids, such as histidine, isoleucine, asparagine, glutamine, threonine and tryptophan, the sequence of consideration of the bonded atoms is important (see Table A.2 in Appendix A). In the current implementation it is ensured by the order of bonds in the list of each atom, determined by the sequence of connections specified in the formula record file (see Table A.3 in Appendix A).

4.2.5   ENHANCEMENT OF COMPUTATIONAL EFFICIENCY

If the system includes a very large number of atoms, even with a cutoff for evaluation of van der Waals interactions, the computation of distances and electrostatic forces between each two atoms A_i and A_j ($i \nsim j$) can become computationally demanding. Therefore a computational mode is implemented, in which the distances and interactions between atoms belonging to distant residues are not considered.

DEFINITION 4.12

Let $\mathcal{N}_i^{[m]}$ be the set of numbers of the main chain atoms belonging to the i-th residue and $\mathcal{N}_i^{[s]} = \mathcal{N}_i \setminus \mathcal{N}_i^{[m]}$. For proline residues we construct $\mathcal{N}_i^{[m]}$ by taking into consideration only atoms typical for the main chain fragments of other residues. Let

$$R_i^{[p]} := \max_{j \in \mathcal{N}_i^{[m]}} \left(\left\| \vec{\mathbf{r}}_{C_\alpha^i j} \right\| + \frac{1}{2} C^{[w]} \right)$$

be called *primary residue radius*, which is the same for non-terminal non-proline residues, and let

$$R_i^{[s]} := \max_{j \in \mathcal{N}_i} \left(\left\| \vec{\mathbf{r}}_{C_\alpha^i j} \right\| + \frac{1}{2} C^{[w]} \right)$$

be referred as *secondary residue radius*.

---

*In PDB format this position is also reserved for atom names

The primary residue radius is permanent, while the secondary residue radius can be reevaluated each time when the residue conformation has changed. As before, $\mathcal{N}_i$ contains the numbers of all atoms in i-th residue, and $C^{[w]}$ denotes the introduced van der Waals cutoff.

In the averaging mode, at first the interactions inside each residue are completely evaluated and the secondary residue radius is determined. The computation of interactions between each two different residues proceeds according to the following rules: if $\|\vec{\mathbf{r}}_{(C_\alpha)_k (C_\alpha)_m}\| > \tilde{k}(R_k^{[s]} + R_m^{[s]})$ (where k, m are residue numbers and $\tilde{k} \geq 1$ is a coefficient), then only the electrostatic interaction between the total residue charges positioned at α-carbons is computed; otherwise the interactions between the side chains are evaluated atomwise. Besides, in the latter case, if $\|\vec{\mathbf{r}}_{(C_\alpha)_k (C_\alpha)_m}\| \leq \tilde{k}(R_k^{[s]} + R_m^{[p]})$, then the interaction of the k-th side chain with the main chain fragment belonging to the m-th residue is evaluated completely, otherwise only the electrostatic interaction of the total charges is computed. Similar procedure is performed for the combination of the main chain fragment of the k-th residue and the m-th side chain. Finally, if $\|\vec{\mathbf{r}}_{(C_\alpha)_k (C_\alpha)_m}\| \leq \tilde{k}(R_k^{[p]} + R_m^{[p]})$, the interaction between the main chain fragments is evaluated atomwise, and it is averaged in the other case. The interaction of the total charges can be replaced by interaction of dipoles, and the scaling constant $\tilde{k}$ can be different for hydrated and non-hydrated residues.

In principle, the strength of electrostatic interactions between hydrated charges decays rather fast with increasing distance (see Subsection 4.3.1). The interaction of the same charges in a protein core is about thirteen times stronger. However, charges tend to stay in contact with water, unless they are sufficiently small to be practically ignored.

4.3   SIMULATIONS

SIVIPROF was designed not only as a protein folding software, but also as a tool for exploration of related phenomena, with an aim of better understanding the folding process and possible model improvement. For example, it may be possible that contributions of some forces are negligible under certain circumstances, or some interactions that are often ignored may be important for correct folding. In the following subsection we shall briefly focus our discussion on these issues and then proceed to energy minimization. Some SIVIPROF simulation results can be also observed on figures in the preceeding sections.

4.3.1   CONTRIBUTIONS OF DIFFERENT INTERACTION TYPES

As discussed in Subsection 1.6.1, van der Waals energy landscapes of main chain fragments help to understand why certain values of dihedral angles prevail in protein structures. A natural question arises: what is the contribution of the electrostatic energy?

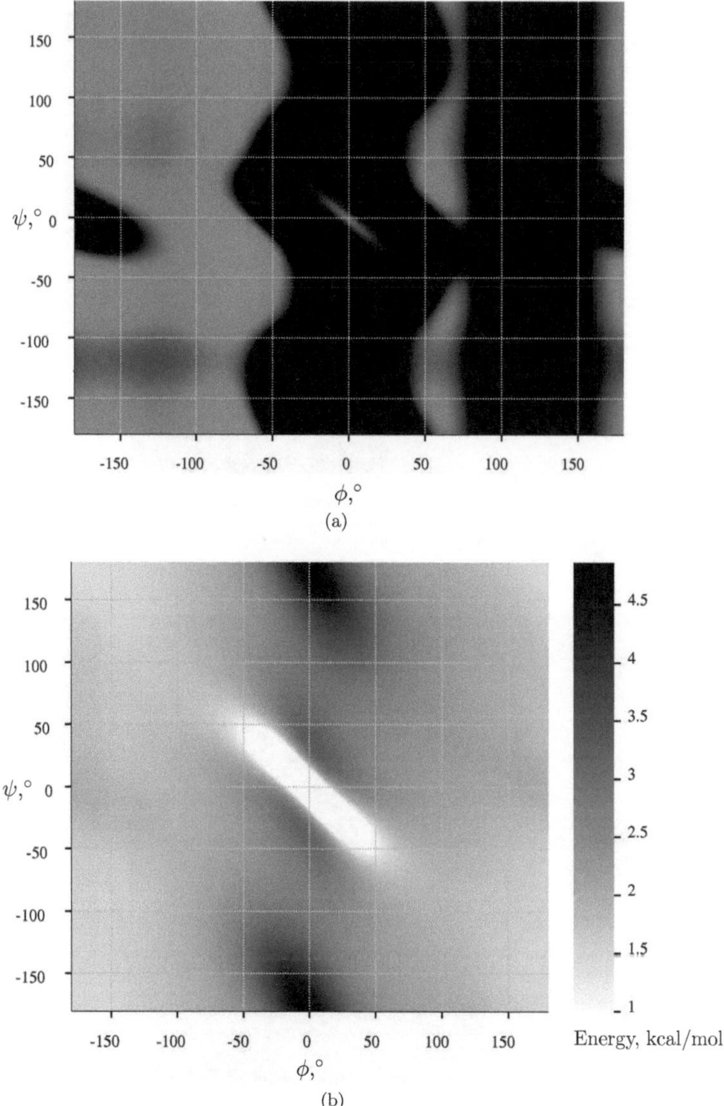

FIG. 4.7: Clipped van der Waals and electrostatic energy surfaces of the peptide fragment shown in Figure 1.26 (a). Black color indicates high energy values. (a) Low values of the van der Waals and electrostatic energy at $\epsilon = 40$ are depicted in green and red respectively. The surfaces are clipped above 0. The minimal value corresponds to -1.76 kcal/mol. (b) The electrostatic energy surface at $\epsilon = 3$. The surface is clipped below 1 kcal/mol.

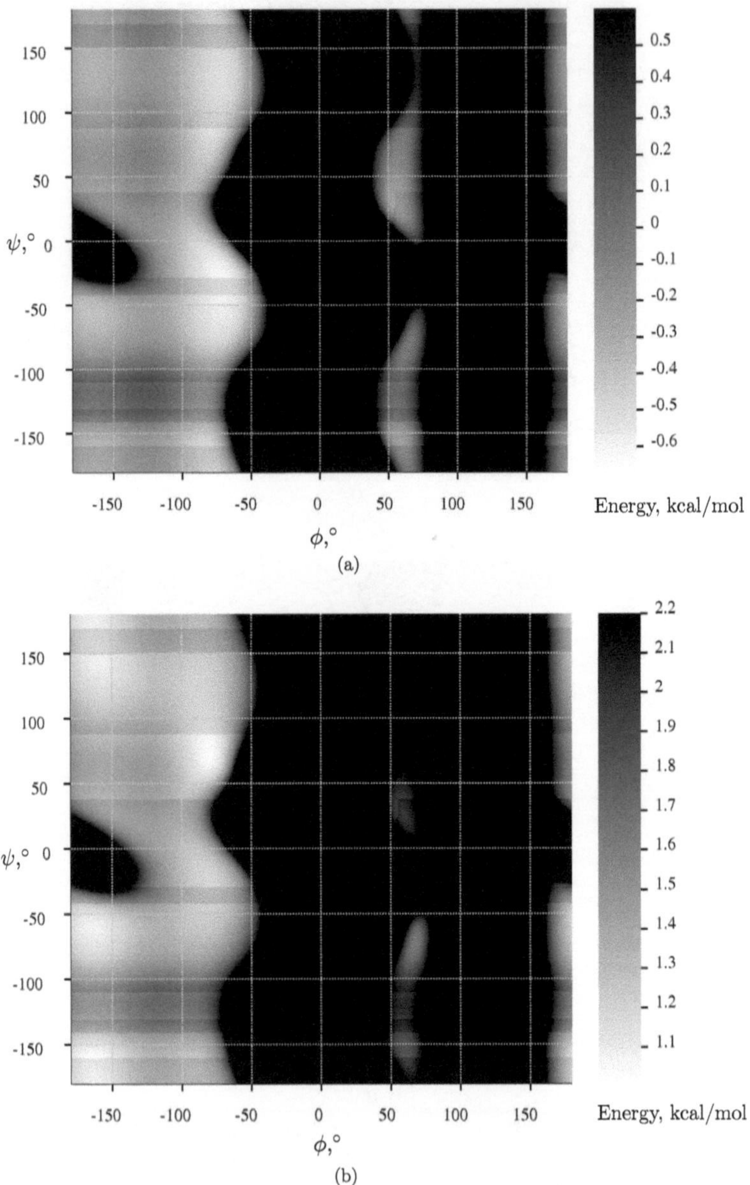

FIG. 4.8: Clipped total energy surfaces of the peptide fragment shown in Figure 1.26 (a). As before, black color indicates high energy values. (a) $\epsilon = 40$. (b) $\epsilon = 3$.

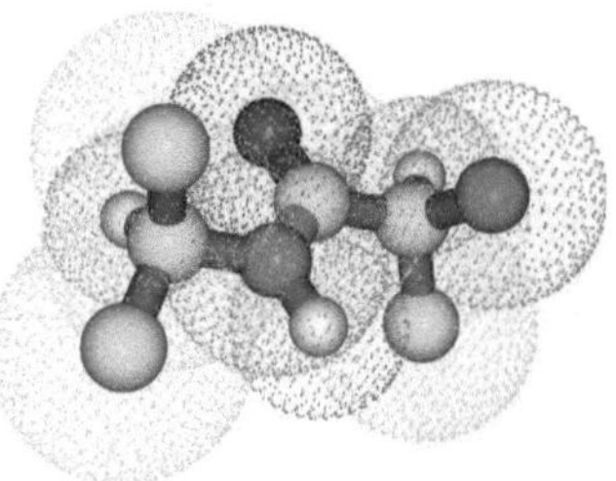

Fig. 4.9: A typical main chain fragment complementary to the one depicted in Figure 1.26 (a).

Figure 4.7 shows color-coded projections of clipped van der Waals and electrostatic energy landscapes, which were obtained for the previously considered typical peptide fragment (see Figure 1.26(a)) with constant electric permittivities. The electrostatic energy depicted in the red color channel in Figure 4.7(a) is computed with $\epsilon = 40$. The minimum in the center appears due to convergence of an oxygen and a hydrogen atom. The valley becomes even better expressed for $\epsilon = 3$ (see Figure 4.7(b)). However, the repulsive van der Waals interaction dominates, therefore the landscape corresponding to the total energy looks similar to the surface considered before (see Figure 4.8). Note, however, that the minimum of the total energy for $\epsilon = 3$ is shifted in the direction of the positive ψ values. The horizontal lines in Figure 4.8 appeared due to solvation energy contributions, which are of no importance for the considered fragment, because some atoms around were removed.

Another question that can be raised in discussion of energy landscapes concerns the interaction of neighboring residues. A fragment that can be regarded as complementary to the one considered above is shown in Figure 4.9. The van der Waals energy landscape for this fragment gives no interesting information, since the bonds $(C_\alpha)_i - (C')_i$ and $(N)_{i+1} - (C_\alpha)_{i+1}$ are parallel and the groups of atoms bound to $(C_\alpha)_i$ and $(C_\alpha)_{i+1}$ do not show significant van der Waals interaction. Even the electrostatic interaction between them for $\epsilon = 40$ is negligible.

However, for $\epsilon = 3$ the interaction between $(N)_i$ and $(C')_{i+1}$ is quite strong (about -1.9 kcal/mol) and must favor a helical conformation over an extended one, in which these atoms have the largest separation. This effect is opposed to the one described above for the total energy in Figure 4.8.

Figure 4.10 (a) shows an exemplary distribution of charges in a tetrapeptide containing hydrophobic and ionized residues. These partial charges are obtained using the method of J. Gasteiger and M. Marsili [26, 27], which is described in Subsection 2.2.3. The other subfigures nearby visualize the resulting forces computed with different electric permittivities. One can see that typical electrostatic forces in vacuum are very strong, while for $\epsilon = 40$ or $\epsilon = 80$ most of them become negligible.

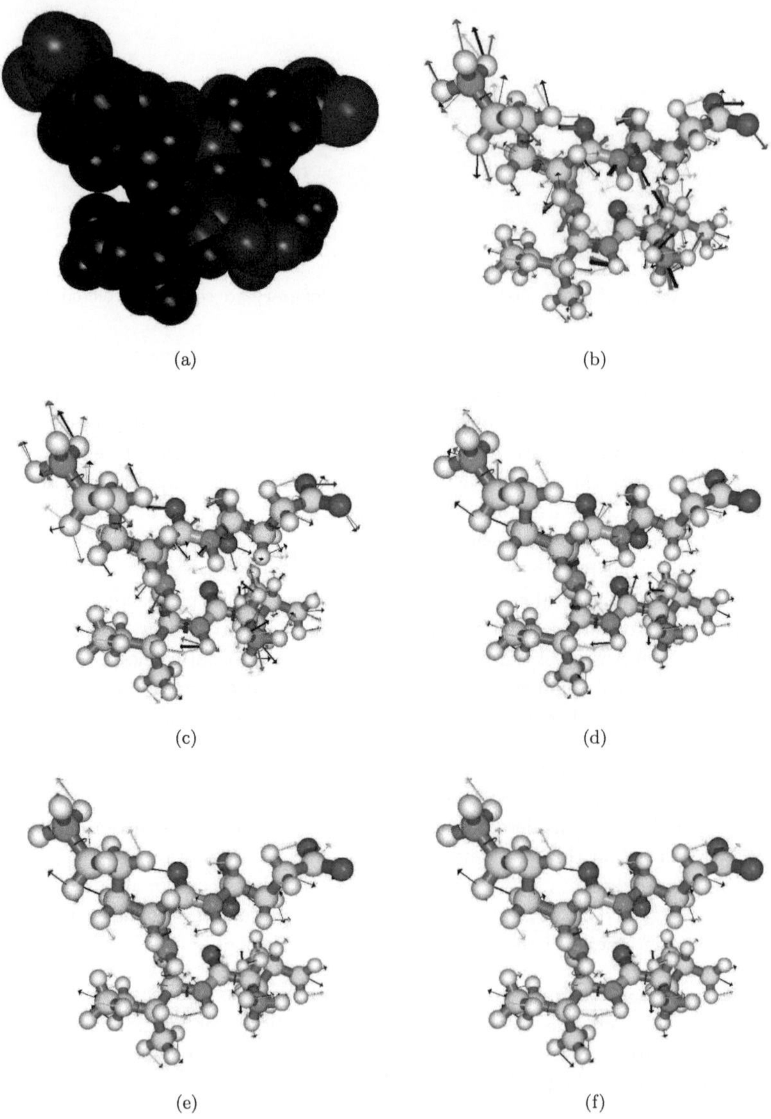

FIG. 4.10: The distribution of charges (a) in the LIKE-tetrapeptide, and the resulting forces computed with different permittivity models (b-f): (b) $\epsilon = 1$, (c) $\epsilon = 3$, (d) the introduced permittivity model, see p. 113, (e) $\epsilon = 40$, (f) $\epsilon = 80$. Black arrows denote total atomic forces. The explanation of other related visualization details is given on p. 31 and on p. 136.

4.3.2 Minimization of energy

As already discussed, we shall expect that the native form of a protein corresponds to a sufficiently deep local minimum of energy. The popular assumption that this minimum should be global and achievable from any initial conformation is not adopted in this work, by the reasons described in Subsections 1.7.3 and 1.7.4. Here, the initial conformation is assumed to be predetermined to some extent by transpeptidation reactions in the active center of a ribosome and possibly even enforced in the ribosomal tunnel. The biochemical basis for this assumption is described in Section 1.8.

However, some tolerance to variations in initial conformations of the side chains, as well as to stochastic collisions with atoms of surrounding solvent at physiological conditions, are presumed. Hence, there is a hope that the system is sufficiently robust to allow predictions even though the natural folding path is not followed exactly, and the model contains significant simplifications. Nevertheless, this speculation suggests that we shall avoid transition paths with unphysiologically high energy barriers. Naturally, this condition is most surely fulfilled if the conformational changes proceed roughly in the direction of twisting forces. Therefore, the choice has been made in favor of the steepest descent and its modifications.

The idea of the steepest descent is to move in the direction of the greatest change downhill on the energy landscape, i.e. opposite to the direction of the gradient. However, since the gradient direction in general changes with the current position, in practice this principle is followed only to a certain extent. Therefore, different realizations of the steepest descent are possible, depending on how far one proceeds in the chosen direction before it will be revised.

One approach consists in performing a line search, in order to locate the next minimum point along the chosen direction, and then update the gradient value to determine a new direction of movement. The procedure is repeated until a minimum is found or a certain stop criterion is fulfilled.

Another way, which is particularly more efficient if energy evaluation is coupled with the computation of the gradient on an analytical basis, is to move from a current point $\vec{\zeta}_k$ using an adaptive step size ε_k, which is increased by some factor (e.g. by 1.2), if the new position,

$$\vec{\zeta}_{k+1} := \vec{\zeta}_k - \frac{\varepsilon_k}{\|\nabla \widetilde{U}(\vec{\zeta}_k)\|} \nabla \widetilde{U}(\vec{\zeta}_k)$$

fulfills a certain criterion. Otherwise the move is rejected, and another attempt is made from the point $\vec{\zeta}_k$ with a reduced step size (e.g., by the factor 0.5 [20]). A typical requirement for a successful step is a decrease in energy $\widetilde{U}$.

Although, strictly speaking, this strategy does not exclude the possibility to miss

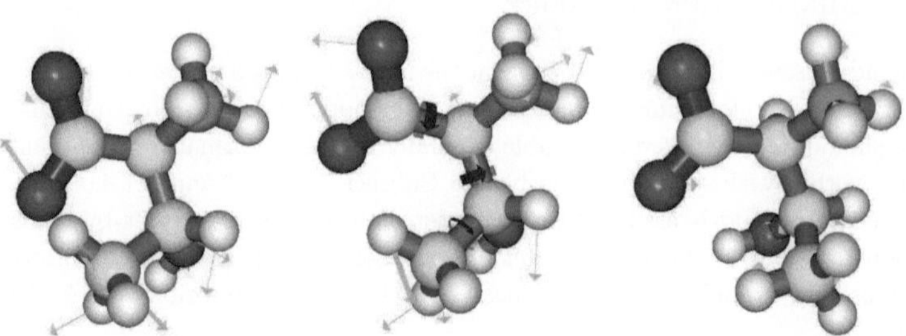

FIG. 4.11: Minimization in Cartesian coordinates compared to minimization in the dihedral angle space. (a) The initial configuration with an intentional distance violation ($\widetilde{U} = 1.4 \times 10^4$ kcal/mol). (b) The configuration after minimization in the space of Cartesian coordinates ($\widetilde{U} = -1.168$ kcal/mol). (c) The configuration after a subsequent minimization in the space of dihedral angles ($\widetilde{U} = -3.59$ kcal/mol).

the "right" minimum by choosing too large steps, practically we can avoid[*] it by setting a reasonable initial value and an appropriate upper bound for the step size.

The steepest descent with an adaptive step size, both in the space of Cartesian coordinates and in the dihedral angle space, in a number of its variations was implemented in SiViProF. Simulations with folding in the space of dihedral angles have shown a significant improvement compared to minimization in Cartesian coordinates, both in efficiency and the quality of the obtained minima.

Figure 4.11 shows the result of minimization in Cartesian coordinates, which was followed by minimization in the dihedral angle space. Minimizations were performed with the same tolerance. However, in the first case the gradient has vanished before the desired precision was achieved[†]. By contrast, if the order of the two minimization types is exchanged, the energy typically does not change so drastically[‡] after the first minimization, and the structural changes are even less noticeable.

The requirement of energy decrease for a successful step the in dihedral angle space works well to resolve distance violations in peptides consisting of several residues (see, for example, Figure 4.1). However, in this case, with exception of very short fragments, minimization finishes in a local minimum often quite close to the initial conformation. That is, the same fragment may accept an α-helical or β-sheet

---

[*]It can be done by taking into account physically meaningful space resolution with respect to the atom size and acceptable fluctuations of energy at the given temperature.

[†]In fact, the gradient has vanished several times during minimization, but the algorithm had some resistance to this event.

[‡]Of course, one can construct examples that would behave differently. However, the minimization in dihedral angle space usually ends with low-energy structures, and subsequent minimization in Cartesian coordinates makes only minor atomic shifts that locally balance the remaining forces by small molecular distortions.

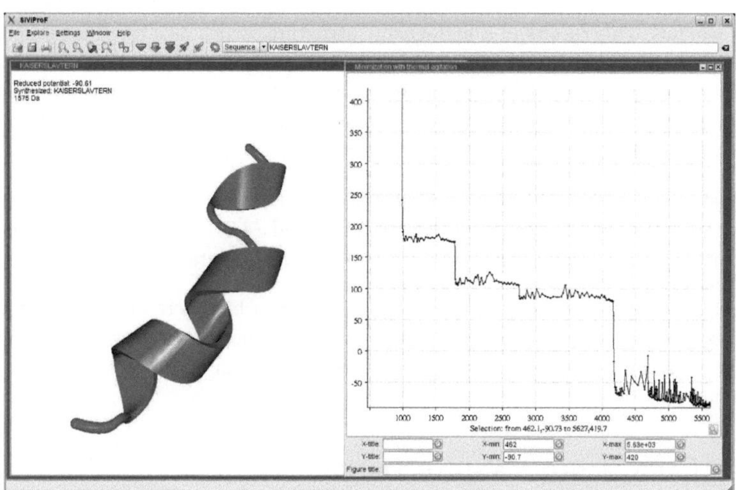

FIG. 4.12: A screenshot of SIVIPROF with results of cotranslational folding simulated using steepest descent allowing energy oscillations (see description in the text).

structure, depending on whether the setup was rather extended or close to helical. Besides, for longer chains the problem becomes stiff, and minimization ends up with high oscillations related to large branches. In these cases even simple bondwise minimization can proceed more successfully.

The described problems are partially resolved by allowing a certain energy increase for a successful transition step. Figure 4.12 shows the result of energy minimization with the following criterion for step acceptance:

$$\widetilde{U}(\vec{\zeta}_{k+1}) - \widetilde{U}(\vec{\zeta}_k) < \bar{R}TM, \quad \text{if } k \leq \tilde{n}, \tag{4.6}$$

or

$$\widetilde{U}(\vec{\zeta}_{k+1}) - \widetilde{U}(\vec{\zeta}_k) \leq \bar{R}T \max(0, M - (k - \tilde{n})), \quad \text{if } k > \tilde{n}. \tag{4.7}$$

Here k and M are, as before, the step number and the number of degrees of freedom, and $\tilde{n}$ is the number of agitated steps, after which the energy decrease is stepwise enforced. In the described simulation, $\tilde{n}$ was set equal to M. The form of the acceptance criterion was inspired by thermodynamical considerations described on p. 42. By contrast, the utilization of the Metropolis Monte Carlo criterion with evaluation of the Boltzmann Factor (see (2.1)) have immediately resulted in complete unfolding and highly non-natural final structures.

The initial conformation for the main chain was preset as suggested by V. I. Lim and A. S. Spirin according to their stereochemistry analysis [43, 72] (see also Subsection 1.8.3 for a more detailed discussion), i.e. both ϕ and ψ were set to $-60°$ in each residue. This results in a high-energy structure, which nevertheless resembles an α-helix. The side chains initially were in an extended conformation. The same fragment of the chain had tendency to unfold in tests with permutations in the sequence, confirming that structural and physical features of residues contribute to the stability of one or another secondary structure.

Simulations with long chains have demonstrated, however, that the described step acceptance criterion, given by conditions (4.6) and (4.7), and the utilized model parameters still require further tuning based on experimental data.

A number of modifications of the gradient direction, which were aimed to decrease stiffness, have been tested in SIVIPROF, but none of them have shown significantly better performance so far. Some further investigations in this direction remain to be done.

4.4   CONCLUSIONS AND OUTLOOK

This work was focused on development of a mathematical model for intracellular protein folding (see Chapter 3) and on implementation of the model in the form of SIVIPROF software for simulation and visualization (described in the current chapter). During the elaboration of the model special attention was given to the factors that may be determinant for physiological folding pathways.

The number of possible conformations of a protein chain is enormous. However, a nascent protein folds in the most cases quickly and reliably into its active form. If a mistake nevertheless occurs, the misfolded protein usually can not refold by itself and requires an intervention of chaperones (see Subsection 1.7.3). Moreover, the probability of failure increases under higher temperature. Besides, an indication that the native form of the protein is not a unique stable state is given by existence of prions (see Subsection 1.7.4). The latter represent a class of proteins that possess, in addition to their native form, another stable structure, which is infectious due to its ability to promote other chains with the same sequence to adopt such a pathogenic fold. Among the diseases caused by prions is the well-known bovine spongiform encephalopathy, often called the mad cow disease. A remarkable feature of the pathogenic form is its exceptional resistance to digestion by enzymes and thermal treatment, challenging the hypothesis about the global minimum of the potential energy, related to native folds.

Therefore, in distinction to other *ab initio* protein folding approaches, which typically suggest that a protein chain can achieve its native state from any initial conformation, a methodology was developed that is aimed to imitate native folding conditions. It assumes that an appropriate initial conformation, favoring certain folding pathways, is enforced on the ribosome and may be necessary for correct folding. The findings from molecular biology that became the basis for this assumption are described in Section 1.8.

To imitate the protein synthesis *in silico**, the algorithms for appropriate coordinate transformations were developed (see Subsection 4.2.2 for generation of initial coordinates, Subsections 1.5.1 and 1.5.2 for the physical basis, and Subsection 3.5.1 for a description of related mathematical operations). The atomic coordinates for amino acids are generated preliminary and then used for subsequent elongation of the desired polypeptide chain. Care is taken about the correct chirality and appropriate ionization states of amino acids (see Sections 1.3 and 1.4 for the background in biochemistry and physical chemistry, and Subsections 4.2.2 and 4.2.3 for the related algorithms). The attachment of a new residue is performed in a way that the formed peptide group is disposed in the *trans* conformation, which prevails significantly in native proteins (see Subsection 1.6.1 for the biochemical basis and Subsection 3.5.2 for the related coordinate transformation). During simulations, the chain is folded subsequently, as when it emerges from a ribosome during protein synthesis. No transitions assuming unphysiological high-energy intermediate states are intentionally allowed (see Subsection 4.3.2). To prevent folding of the nascent chain about the region of the chain elongation, a special energy term is incorporated that enforces the preference to a desired half-space (see Section 3.5).

The force field was formulated in the space of dihedral angles (see Section 3.3) to enable more efficient energy minimization. For this purpose, some analysis of relations between the interatomic distances and intramolecular dihedral angles was performed (see Section 3.2). A further improvement of efficiency was achieved by computation of analytical derivatives, which are related to atomic and twisting forces (see Section 3.6). In particular, the forces arising from the van der Waals and electrostatic interactions (see Subsections 1.5.3 and 1.5.4 for the physical background), as well as from bond stretching and angle bending in disulfide bridges (see Sections 1.2, 1.4, and Subsection 1.8.4 for the biochemical background), were discussed in this context.

Beside the effects described above, the interaction with the surrounding solvent is crucial in determination of the final protein structure (see Subsections 1.7.1 and 1.7.2). In particular, the hydrophobic effect, which has largely entropic nature under physiological conditions, is regarded by biochemists as the major driving force in the process of protein folding. Since explicit inclusion of water molecules into simulations is very expensive from the computational point of view, an implicit solvation model was developed with an aim to reproduce the desired effects (see Section 3.4). The dynamics of protein atoms in solution was analyzed in Section 3.7. The conclusion was made, that an appropriate scaling of the gradient may help to obtain a transformation path better describing a natural folding pathway.

By reasons motivating the character of the model, the energy minimization methods were selected with attention to the nature of the described processes. On the other hand, the computational efficiency is an important issue particularly for this prob-

---

*I.e., via computer simulations.

lem. Although full-time molecular dynamics involving solvent atoms (and possibly with consideration of quantum effects) would be the most appropriate to reconstruct the folding process in all details, it is not feasible in the most cases, and reasonable simplifications are necessary.

The steepest descent in the space of dihedral angles appeared to be a natural choice for the proposed model. This method performs significantly better than different energy minimization methods carried out in Cartesian coordinates. The latter, namely, tend to resolve high energy configurations mainly by angle bending. However, for long chains, the steepest descent in the dihedral angle space ends up with high oscillations. Besides, the requirement of energy decrease for successive conformational transitions may be unphysical, since naturally happening stochastic collisions with solvent atoms are known to be the driving force in many conformational transitions and chemical reactions. Therefore, the step acceptance criterion for this method was modified to allow certain increase in energy. Other methods have been tested also, but none of them have yielded better results.

Determination of the optimal parameter sets by comparison of simulation predictions with native structures, as well as the related refinement of the model still remains to be done. Particularly the solvation parameters require verification, which is difficult in the light of the large discrepancy in experimental values. Further work shall be directed to the revision of the energy minimization method with involvement of statistical thermodynamics. The evaluation of the electrostatic interactions can be possibly improved by introducing the electrical permittivity depending on the distance from the solvent exposed surface. Models reconstructing the ionization behavior and disulfide bond reshuffling, as well as *cis-trans* isomerization of peptide groups with proline residues can be included into simulations.

Since some new biochemical findings suggest that ribosomes may enforce different initial conformations for specific sequences (see Subsection 1.8.3), it may be worth trying to imitate this effect by using secondary structure predictions from knowledge-based methods for selective setup of the initial main chain configuration in simulations with cotranslational folding. The possible impact of initial side chain conformations shall be explored also. The tools necessary for the latter investigations are already implemented in SIVIPROF.

Implementation Details

A.1 Visual representation of molecular elements

Table A.1: Representation of atoms and bonds on molecular images and the corresponding symbols used for program input.

Representation	Symbols	Object
	C	Carbon atom
	H	Hydrogen atom
	N	Nitrogen atom
	O	Oxygen atom
	S	Sulfur atom
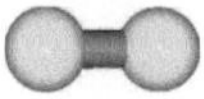	–	Single bond
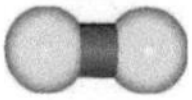	=	Double bond
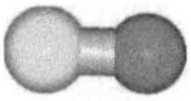	+	Peptide bond
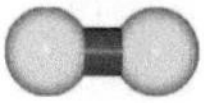	*	Aromatic or partially double non-peptide bond
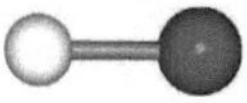		Hydrogen bond

A.2 ATOM NUMERATION IN AMINO ACIDS

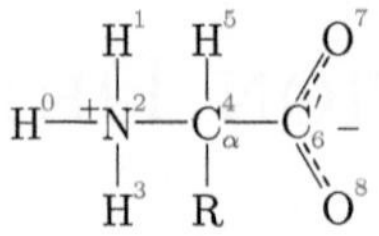

$$H^0 \!-\! {}^{\pm}N^2 \!-\! C^4_\alpha \!-\! C_6 \quad\quad\quad (H^1,\ H^3,\ H^5,\ R,\ O^7,\ O^8)$$

FIG. A.1: The numeration of atoms in the common part of amino acids. Atom charges given here and in the following table are formal. They are assigned as initial charges for the net charge computation.

TABLE A.2: Common abbreviations for standard amino acids, atom numbering used in SIVIPROF (red indices) and conventional atomic identifiers (blue labels).

One-letter code	Three-letter code	Name	Side chain numbering
A	ala	alanine	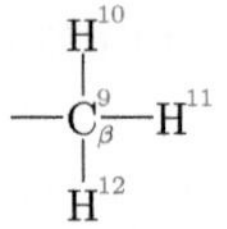 $-C^9_\beta$ with H^{10}, H^{11}, H^{12}
C	cys	cysteine	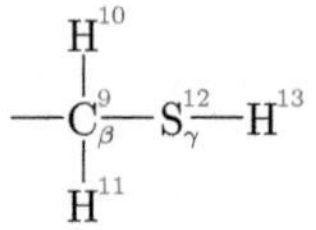 $-C^9_\beta$ with H^{10}, H^{11}, $-S^{12}_\gamma\!-\!H^{13}$
D	asp	aspartic acid	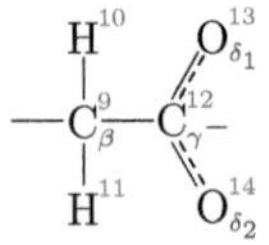 $-C^9_\beta$ with H^{10}, H^{11}, $-C^{12}_\gamma\!-$ with $O^{13}_{\delta_1}$, $O^{14}_{\delta_2}$
E	glu	glutamic acid	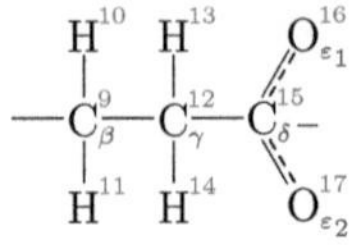 $-C^9_\beta$ (H^{10}, H^{11}) $-C^{12}_\gamma$ (H^{13}, H^{14}) $-C^{15}_\delta\!-$ with $O^{16}_{\varepsilon_1}$, $O^{17}_{\varepsilon_2}$
F	phe	phenylalanine	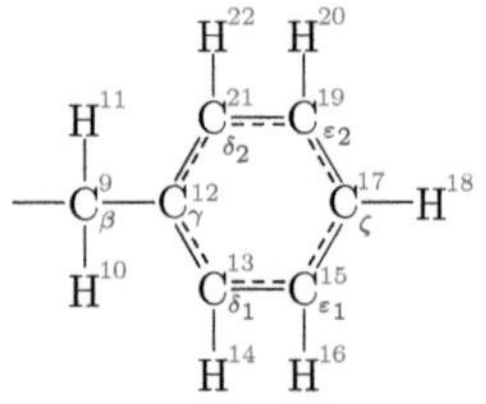 $-C^9_\beta$ (H^{10}, H^{11}) $-C^{12}_\gamma$; ring: $C^{21}_{\delta_2}$ (H^{22}), $C^{19}_{\varepsilon_2}$ (H^{20}), $C^{17}_\zeta\!-\!H^{18}$, $C^{13}_{\delta_1}$ (H^{14}), $C^{15}_{\varepsilon_1}$ (H^{16})
G	gly	glycine	$-H^9$

continued on the next page

TABLE A.2 *(continued)*

One-letter code	Three-letter code	Name	Side chain numbering
H	his	histidine	H^{10}—C^9_β—H^{11}; C^{12}_γ = $N^{18}_{\delta_1}$—H^{19}; $C^{13}_{\delta_2}$, H^{14}; $N^{15}_{\varepsilon_2}$; $C^{16}_{\varepsilon_1}$—H^{17}
I	ile	isoleucine	H^{12}—$C^{11}_{\gamma_2}$—H^{14}, H^{13}, H^{10}; C^9_β—$C^{15}_{\gamma_1}$—C^{18}_δ—H^{20}; H^{19}, H^{16}, H^{17}, H^{21}
K	lys	lysine	H^{10}, H^{11}—C^9_β—C^{12}_γ(H^{13},H^{14})—C^{15}_δ(H^{16},H^{17})—C^{18}_ε(H^{19},H^{20})—N^{21+}_ζ(H^{22},H^{24})—H^{23}
L	leu	leucine	H^{16}, H^{15}—$C^{14}_{\delta_1}$—H^{17}; C^9_β—C^{12}_γ—$C^{18}_{\delta_2}$—H^{21}; H^{10},H^{11},H^{13},H^{19},H^{20}
M	met	methionine	H^{10},H^{11}—C^9_β—C^{12}_γ(H^{13},H^{14})—S^{15}_δ—C^{16}_ε(H^{17},H^{19})—H^{18}
N	asn	asparagine	H^{10},H^{11}—C^9_β—C^{12}_γ = $O^{13}_{\delta_1}$; $N^{14}_{\delta_2}$—H^{15}, H^{16}
P	pro	proline	H^{14}, H^{13}—C^{12}_γ—C^9_β(H^{11},H^{10}); H^{16},H^3—C^{15}_δ—N^{2+}(H^0,H^1)—C^4_α(H^5)—C^6 = O^8, O^7_-

continued on the next page

Table A.2 *(continued)*

One-letter code	Three-letter code	Name	Side chain numbering
Q	gln	glutamine	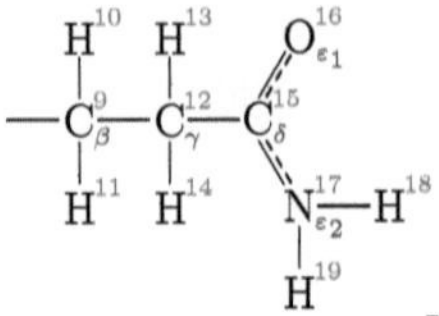 H^{10}, H^{13}, $O^{16}_{\varepsilon_1}$; $-C^{9}_{\beta}-C^{12}_{\gamma}-C^{15}_{\delta}$; H^{11}, H^{14}, $N^{17}_{\varepsilon_2}-H^{18}$, H^{19}
R	arg	arginine	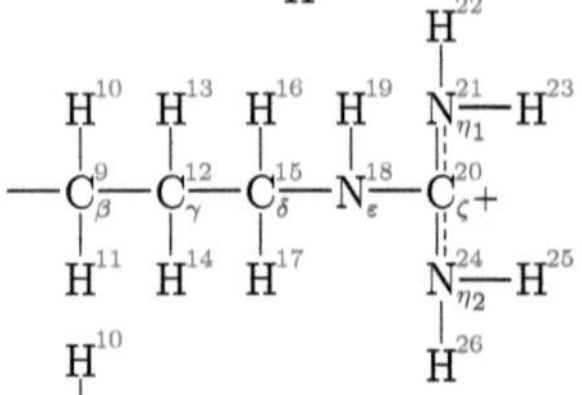 H^{10}, H^{13}, H^{16}, H^{19}, $N^{21}_{\eta_1}-H^{23}$, H^{22}; $-C^{9}_{\beta}-C^{12}_{\gamma}-C^{15}_{\delta}-N^{18}_{\varepsilon}-C^{20}_{\zeta}+$; H^{11}, H^{14}, H^{17}, $N^{24}_{\eta_2}-H^{25}$, H^{26}
S	ser	serine	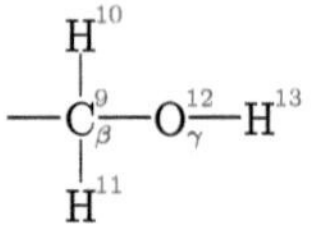 H^{10}; $-C^{9}_{\beta}-O^{12}_{\gamma}-H^{13}$; H^{11}
T	thr	threonine	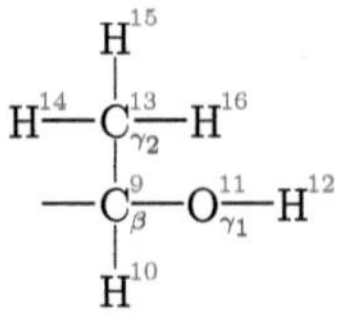 H^{15}; $H^{14}-C^{13}_{\gamma_2}-H^{16}$; $-C^{9}_{\beta}-O^{11}_{\gamma_1}-H^{12}$; H^{10}
V	val	valine	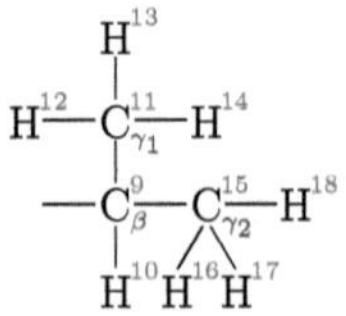 H^{13}; $H^{12}-C^{11}_{\gamma_1}-H^{14}$; $-C^{9}_{\beta}-C^{15}_{\gamma_2}-H^{18}$; H^{10}, H^{16}, H^{17}
W	trp	tryptophan	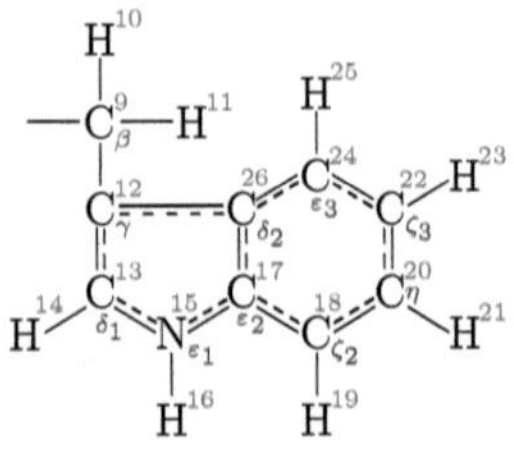 H^{10}; $-C^{9}_{\beta}-H^{11}$; C^{12}_{γ}, $C^{26}_{\delta_2}$, $C^{24}_{\varepsilon_3}$, H^{25}, $C^{22}_{\zeta_3}-H^{23}$; $C^{13}_{\delta_1}$, H^{14}, $N^{15}_{\varepsilon_1}$, $C^{17}_{\varepsilon_2}$, $C^{18}_{\zeta_2}$, $C^{20}_{\eta}-H^{21}$, H^{16}, H^{19}
Y	tyr	tyrosine	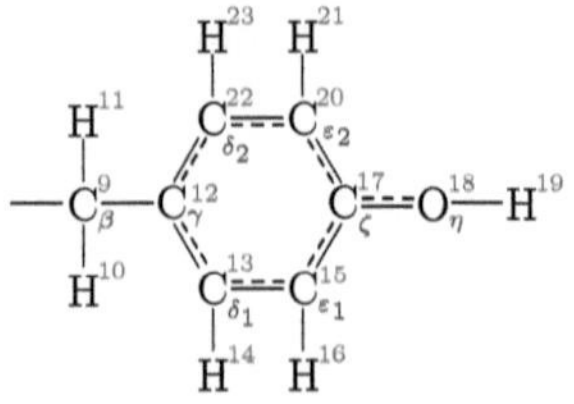 H^{11}; $-C^{9}_{\beta}-C^{12}_{\gamma}$; $C^{22}_{\delta_2}-H^{23}$, $C^{20}_{\varepsilon_2}-H^{21}$, $C^{17}_{\zeta}-O^{18}_{\eta}-H^{19}$; H^{10}, $C^{13}_{\delta_1}-H^{14}$, $C^{15}_{\varepsilon_1}-H^{16}$

A.3 SiViProF input format

SiViProF can create molecules of various chemicals by name, if the description of the corresponding chemical structure is available in one of the considered input files. There are two input files, one for amino acids, and the other one for the rest of chemicals. Each file consists of line records, containing the name, the internal format formula and the bonding hint for a certain chemical, which are separated by tabulation. Records for amino acids describe only side chains and include one- and tree-letter codes as well. A record for the common part is given separately.

The internal format formula contains consequently merged atom records consisting of the type of the element, the number of bonds and the formal charge, if it differs from zero. Specification of the number of bonds is used for convenience in allocation of memory. Bonding hint is build of a sequence of bond records, separated by comma and space. Each bond record consists of a couple of atom numbers, separated by a bond sign (see Table A.1).

Only creation of molecules, containing chemical elements in hybridization states that are represented in amino acids, is currently implemented. No support of structures other then non-cyclic and containing single or double rings is guaranteed.

TABLE A.3: Formulas of amino acids in SiViProF input format.

Name	SiViProF formula and bonding hint
Common part	H1H1+N4H1C4H1-C3O1O1 0-2, 1-2, 2-3, 2-4, 4-5, 4-6, 6*7, 6*8
Alanine	C4H1H1H1 9-10, 9-11, 9-12
Arginine	C4H1H1C4H1H1C4H1H1N3H1+C3N3H1H1N3H1H1 9-10, 9-11, 9-12, 12-13, 12-14, 12-15, 15-16, 15-17, 15-18, 18-19, 18-20, 20*21, 21-22, 21-23, 20*24, 24-25, 24-26
Asparagine	C4H1H1C3O1N3H1H1 9-10, 9-11, 9-12, 12*13, 12*14, 15-14, 14-16
Aspartic acid	C4H1H1-C3O1O1 9-10, 9-11, 9-12, 12*13, 12*14
Cysteine	C4H1H1S2H1 9-10, 9-11, 9-12, 12-13
Glutamic acid	C4H1H1C4H1H1-C3O1O1 9-10, 9-11, 9-12, 12-13, 12-14, 12-15, 15*16, 15*17
Glutamine	C4H1H1C4H1H1C3O1N3H1H1 9-10, 9-11, 9-12, 12-13, 12-14, 12-15, 15*16, 15*17, 17-18, 17-19

continued on the next page

TABLE A.3 (continued)

Name	SiViProF formula and bonding hint
Glycine	H1
Histidine	C4H1H1C3C3H1N2C3H1N3H1 9-10, 9-11, 9-12, 12*18, 12*13, 13-14, 13*15, 15*16, 16-17, 16*18, 18-19
Isoleucine	C4H1C4H1H1H1C4H1H1C4H1H1H1 9-15, 9-10, 9-11, 11-12, 11-13, 11-14, 15-16, 15-17, 15-18, 18-19, 18-20, 18-21
Leucine	C4H1H1C4H1C4H1H1H1C4H1H1H1 9-10, 9-11, 9-12, 12-13, 12-14, 12-18, 14-15, 14-16, 14-17, 18-19, 18-20, 18-21
Lysine	C4H1H1C4H1H1C4H1H1C4H1H1+N4H1H1H1 9-10, 9-11, 9-12, 12-13, 12-14, 12-15, 15-16, 15-17, 15-18, 18-19, 18-20, 18-21, 21-22, 21-23, 21-24
Methionine	C4H1H1C4H1H1S2C4H1H1H1 9-10, 9-11, 9-12, 12-13, 12-14, 12-15, 15-16, 16-17, 16-18, 16-19
Phenylalanine	C4H1H1C3C3H1C3H1C3H1C3H1C3H1 9-10, 9-11, 9-12, 12*13, 13-14, 15*13, 15-16, 15*17, 17-18, 19*17, 19-20, 19*21, 21-22, 21*12
Proline	H1H1+N4H1C4H1-C3O1O1C4H1H1C4H1H1C4H1 0-2, 1-2, 2-4, 2-15, 4-5, 4-6, 4-9, 9-10, 9-11, 9-12, 12-13, 12-14, 12-15, 15-16, 15-3, 6*7, 6*8
Threonine	C4H1O2H1C4H1H1H1 9-10, 9-11, 11-12, 9-13, 13-14, 13-15, 13-16
Tryptophan	C4H1H1C3C3H1N3H1C3C3H1C3H1C3H1C3H1C3 9-10, 9-11, 9-12, 12*13, 13-14, 13*15, 15*17, 16-15, 18*17, 18-19, 18*20, 20-21, 22*20, 22-23, 22*24, 24-25, 26*17, 26*24, 26*12
Tyrosine	C4H1H1C3C3H1C3H1C3O2H1C3H1C3H1 9-10, 9-11, 9-12, 12*13, 13-14, 15*13, 15-16, 15*17, 17*18, 18-19, 20*17, 21-20, 20*22, 23-22, 22*12
Serine	C4H1H1O2H1 9-10, 9-11, 9-12, 12-13
Valine	C4H1C4H1H1H1C4H1H1H1 9-10, 9-11, 9-15, 11-12, 11-13, 11-14, 15-16, 15-17, 15-18

APPENDIX B

PHYSICAL QUANTITIES, CONSTANTS AND UNITS

TABLE B.1: The base SI units [14], p. 960.

Name	Symbol	Physical quantity
ampere	A	electric current
candela	cd	luminous intensity
kelvin	K	thermodynamic temperature
kilogram	kg	mass
meter	m	length
mole	mol	amount of substance
second	s	time

TABLE B.2: Units used in this work and related physical quantities.

Name	Symbol	Conversion to base SI units	Physical quantity
ångström	Å	10^{-10} m	length
elementary charge	$|e^-|$	1.6022×10^{-19} A $\times$ s	electric charge
debye	D	3.3356×10^{-30} A $\times$ s $\times$ m	dipole moment
kilocalorie	kcal	4.1868×10^3 kg $\times$ m^2 $\times$ s^{-2}	energy
dalton	Da	1.6605×10^{-27} kg	mass

TABLE B.3: Relevant physical constants.

Name	Symbol	Value
Avogadro's number	N_A	6.0221×10^{23} mol^{-1}
vacuum permittivity	ϵ_0	1.4441×10^{20} $|e^-|^2 \times (\text{Å} \times \text{kcal})^{-1}$
Boltzmann's constant	k_B	3.2975×10^{-27} kcal $\times$ K^{-1}
gas constant	$\bar{R}$	1.9859×10^{-3} kcal $\times (\text{K} \times \text{mol})^{-1}$
Planck's constant	h	1.5826×10^{-37} kcal $\times$ s
reduced Planck's constant	$\hbar$	2.5188×10^{-38} kcal $\times$ s

APPENDIX C

MATHEMATICAL NOTATIONS

C.1   GENERAL NOTATIONS

$i = \overline{1,n}$	$i \in \{1, 2, \ldots, n\}$
$\overline{\Psi}$	complex conjugate of Ψ
$\wedge$	logical AND
$\vee$	logical OR
∇	gradient
Δ	Laplace operator, $\Delta := \frac{\partial^2}{\partial x^2} + \frac{\partial^2}{\partial y^2} + \frac{\partial^2}{\partial z^2}$
Δx	increment of x
$\|\vec{v}\|$	Euclidean norm of $\vec{v}$
$\vec{a} \cdot \vec{b}$	scalar product of $\vec{a}$ and $\vec{b}$
$\vec{a} \times \vec{b}$	vector product of $\vec{a}$ and $\vec{b}$
$\times$	multiplication sign, e.g. utilized between numbers or units
$\circ$	function composition
$\angle(\vec{a}, \vec{b})$	angle between $\vec{a}$ and $\vec{b}$
$\mathbf{A}^{\mathrm{T}}$	transpose of $\mathbf{A}$
$\operatorname{card} \mathcal{L}$	cardinality of $\mathcal{L}$
$\mathcal{A} \cap \mathcal{B}$	set intersection
$\mathcal{A} \cup \mathcal{B}$	set union
$\mathcal{A} \setminus \mathcal{B}$	set difference

C.2   GENERAL EXPRESSIONS

$\forall$	"for all"
$\nexists$	"does not exist"
$\doteq$	"must be equal"
$\vec{a} \nparallel \vec{b}$	"$\vec{a}$ and $\vec{b}$ are non-collinear"

C.3 Problem-specific notations

$\mathcal{N}$ — set of atom numbers

N — $\operatorname{card}\mathcal{N}$

A_i — i-th atom

$R_i^{[w]}$ — van der Waals radius of A_i

q_i — partial charge of A_i

$\vec{\mathbf{r}}_i$ — position of A_i

$\vec{\mathbf{r}}_{ij}$ — $\vec{\mathbf{r}}_j - \vec{\mathbf{r}}_i$

ζ_{jk}^{il} — dihedral angle $\angle(A_i - A_j - A_k - A_l)$, also: $\angle(\vec{\mathbf{r}}_{ij}, \vec{\mathbf{r}}_{jk}, \vec{\mathbf{r}}_{kl})$

$i \bowtie j$ — A_i and A_j are covalently bound

$i \nsim j$ — A_i and A_j are not covalently bound, and $\nexists k \in \mathcal{N}: \ i \bowtie k \bowtie j$

$i \simeq j$ — A_i and A_j are joined by a double or partially double bond

b_{ij} — directional bond of A_i to A_j, see p. 109

w_{ij} — weight of b_{ij}, see p. 110

$\widehat{\mathcal{R}}_{ij}$ — union of all rings containing b_{ij}, see p. 110

$\mathcal{D}_{ij}$ — branch originated by b_{ij}, see p. 110

$\mathcal{N}_{jk}$ — $\{n \in \mathcal{N} \mid A_n \in \mathcal{D}_{jk}\}$

d_{ij} — distance between A_i and A_j as a function of torsion angles, p. 111

l_{ij} — optimal length of b_{ij}

α_{ij} — optimal bond angle in A_i

$\vec{\zeta}$ — vector of variable primary dihedral angles (list of degrees of freedom)

$\mathcal{M}$ — indices of degrees of freedom

M — $\operatorname{card}\mathcal{M}$

$\kappa(i,j)$ — index of b_{ij} in the list of degrees of freedom, see p. 110

$\iota(i)$ — bond/branch indices related to i-th degree of freedom, see p. 133

$\aleph_{ij}$ — minimal number of consequently bonded atoms between A_i and A_j

More notations are introduced locally.

C.4 Introduced operations

$\vec{\mathbf{a}}\!\downarrow\!\alpha$ — deflection of vector $\vec{\mathbf{a}}$ to angle α

$\vec{\mathbf{a}}\!\downarrow^{\vec{\mathbf{b}}}\!\alpha$ — deflection of vector $\vec{\mathbf{a}}$ to angle α in the direction of vector $\vec{\mathbf{b}}$

$\vec{\mathbf{a}}\,{}_\vee^\circ\,\vec{\mathbf{b}}$ — deflection of vector $\vec{\mathbf{a}}$ in the direction of vector $\vec{\mathbf{b}}$ to match angle α

$\mathbf{M}_{\vec{\mathbf{v}}}(\alpha)\vec{\mathbf{a}}$ — rotation of vector $\vec{\mathbf{a}}$ to angle α about vector $\vec{\mathbf{v}}$

$\mathbf{M}_{\vec{\mathbf{b}}}^{\vec{\mathbf{a}}}\vec{\mathbf{a}}$ — matching the direction of vector $\vec{\mathbf{a}}$ with vector $\vec{\mathbf{b}}$

$\mathbf{T}_{\vec{\mathbf{b}}}^{\vec{\mathbf{a}}}\vec{\mathbf{c}}$ — specific coordinate transformation, see the description on p. 126

$\Xi_{\mathcal{B}}(\vec{\mathbf{v}})$ — coordinate representation of $\vec{\mathbf{v}}$ in basis $\mathcal{B}$

$\Upsilon(r, \varphi, \vartheta)$ — mapping spherical to Cartesian coordinates

C.5 CONVENTIONAL NOTATIONS

$$\phi_i \qquad \angle(C'_{i-1}-N_i-(C_\alpha)_i-C'_i)$$
$$\psi_i \qquad \angle(N_i-(C_\alpha)_i-C'_i-N_{i+1})$$
$$\omega_i \qquad \angle((C_\alpha)_i-C'_i-N_{i+1}-(C_\alpha)_{i+1})$$

$\chi_i^1,\ \chi_i^2,\ \ldots$ side chain dihedral angles, starting from the bond to C_α

PARAMETERS

TABLE D.1: van der Waals radii $R^{[w]}$ and Lennard-Jones potential parameters $K^{[w]}$, adopted from [124] and [94] respectively*.

Chemical element	$R^{[w]}$, Å	$K^{[w]}$, kcal/mol
C	1.70	0.107
H	1.20	0.042
N	1.55	0.095
O	1.52	0.116
S	1.80	0.314

TABLE D.2: Parameters for computation of Gasteiger partial charges, inferred from [27].

A_i element and specification	a_i	b_i	c_i
C, sp^3	7.98	9.18	1.88
C, sp^2	8.79	9.32	1.51
N, sp^2	12.87	11.15	0.85
N, cationic	11.54	10.82	1.36
O, sp^3	14.18	12.92	1.39
O, sp^2	17.07	13.79	0.47
H	7.17	6.24	−0.56
S, sp^3	10.14	9.13	1.38

Atoms with partially double bonds are treated as sp^2-hybridized.

---

*The van der Waals radii specified in [94] coincide with the ones from [124] for protein atoms.

TABLE D.3: Bond stretching parameters, adopted from [94].

A_i element and specification	A_j element and specification	Bond*	l_{ij}, Å	$k_{ij}^{[b]}, \frac{\text{kcal}}{\text{mol} \times \text{Å}^2}$
C, sp^3	C, sp^3	−	1.54	633.6
C, sp^3	C, aromatic	−	1.525	640
C, sp^2	C, sp^3	−	1.501	639
C, sp^2	C, sp^2	=	1.335	1340
C, sp^2	C, aromatic	−	1.51	1340
C, aromatic	C, aromatic	*	1.395	1400
C, sp^3	H	−	1.1	622.4
C, sp^2	H	−	1.089	692
C, aromatic	H	−	1.084	692
C, sp^3	N, cationic	−	1.47	760
C, sp^2	N, cationic	−	1.33	1300
C, sp^3	N, in peptide group	−	1.45	677.6
C, sp^2	N, sp^2	−	1.444	1300
C, sp^2	N, sp^2	=	1.27	1305.94
C, sp^2	N, in peptide group	+	1.345	870.1
C, sp^2	N, planar†	−	1.3	1200
C, aromatic	N, planar†	−	1.35	1306
C, sp^3	O, sp^3	=	1.43	618.9
C, sp^2	O, sp^3	=	1.33	699.84
C, sp^2	O, sp^2	=	1.22	1555.2
C, in carboxyl group	O, in carboxyl group	*	1.24	699.84
C, aromatic	O, sp^3	−	1.39	700
C, sp^3	S, sp^3	−	1.817	381.6
H	N, cationic	−	1.08	692
H	N, in peptide group	−	1	700
H	N, planar†	−	1.03	692
H	O, sp^3	−	0.95	1700.5
H	S, sp^3	−	1.008	700

*See Table A.1 for bond type notations.

†Nitrogen, that has no double bonds, but sp^2-hybridized due to delocalization of electrons.

TABLE D.4: Reference bond angles and angle bending parameters, inferred from [94] .

A_i element and specification	$\alpha_i,^\circ$	$k_i^{[a]}, \dfrac{\text{kcal}}{\text{mol} \times (^\circ)^2}$
C, sp^3	109.5	0.02
C, sp^2	120	0.024
C, aromatic	120	0.024
N, sp^2	120	0.04
N, planar	120	0.04
N, in peptide group	120	0.02
N, cationic	109.5	0.01
O, sp^3	109.5	0.02
S, sp^3	97	0.02

Solvation parameters are computed in accordance with the information summarized in Subsections 1.7.1 and 1.7.2. Atoms with the partial charges less then $0.2\,|e^-|$ are currently treated as hydrophobic. However, this requires further experimental verification.

REFERENCES

[1] A. V. FINKELSTEIN AND O. B. PTIZIN, *Protein physics*, Academic Press, an Imprint of Elsevier Science, 2002.

[2] T. NOGUTI AND N. GŌ, *A method for rapid calculation of a second derivative matrix of conformational energy for large molecules*, J. Phys. Soc. Japan, 52 (1983), pp. 3685–3690.

[3] K. D. GIBSON AND H. A. SCHERAGA, *Dynamics of peptides with fixed geometry: kinetic energy terms and potential energy derivatives as function of dihedral angles*, J. Comput. Chem., 11 (1990), pp. 487–492.

[4] P. KOEHL, J.-F. LEFÈVRE, AND O. JARDETZKY, *Computing the geometry of a molecule in dihedral angle space using n.m.r.-derived constrains. A new algorithm based on optimal filtering*, J. Mol. Biol., 223 (1992), pp. 299–315.

[5] A. NEUMAIER, *Molecular modeling of proteins and mathematical prediction of protein structure*, SIAM Rev., 39 (1997), pp. 407–460.

[6] H. BAUMANN, K. PAULSEN, H. KOVÁCS, H. BERGLUND, A. P. H. WRIGHT, J.-A. GUSTAFSSON, AND T. HÄRD, *Refined solution structure of the glucocorticoid receptor DNA-binding domain*, Biochemistry, 32 (1993), pp. 13463–13471.

[7] J. SANTORO, C. GONZÁLEZ, M. BRUIX, J. L. NEIRA, J. L. NIETO, J. HERRANZ, AND M. RICO, *High-resolution three-dimensional structure of ribonuclease A in solution by nuclear magnetic resonance spectroscopy*, J. Mol. Biol., 229 (1993), pp. 722–734.

[8] C. B. ANFINSEN AND E. HABER, *Studies on the reduction and re-formation of protein disulfide bonds*, J. Biol. Chem., 236 (1961), pp. 1361–1363.

[9] A. JILEK, C. MOLLAY, C. TIPPELT, J. GRASSI, G. MIGNOGNA, J. MÜLLEGGER, V. SANDER, C. FEHRER, D. BARRA, AND G. KREIL, *Biosynthesis of a D-amino acid in peptide linkage by an enzyme from frog skin secretions*, Proc. Natl. Acad. Sci USA, 102 (2005), pp. 4235–4239.

[10] E. C. JIMENÉZ, B. M. OLIVERA, W. R. GRAY, AND L. J. CRUZ, *Contryphan is a D-tryptophan-containing Conus peptide*, J. Biol. Chem., 271 (1996), pp. 28002–28005.

[11] K. PISAREWICZ, D. MORA, F. PFLUEGER, G. FIELDS, AND F. MARI, *Polypeptide chains containing D-gamma-hydroxyvaline*, J. Am. Chem. Soc., 127 (2005), pp. 6207–6215.

[12] S. BARON AND ED., *Medical microbiology*, Galveston (TX): University of Texas Medical Branch, 1996.

[13] P. WAAGE AND C. M. GULDBERG, *Études sur les affinités chimiques*, Forhandlinger: Videnskabs-Selskabet i Christiana, 1864.

[14] P. ATKINS AND J. DE PAULA, *Atkins' physical chemistry*, Oxford University Press, 2006.

[15] D. MESCHEDE, *Gerthsen Physik*, Springer, Berlin, 2004.

[16] H. HEUSER, *Lehrbuch der Analysis, Teil 1*, B. G. Teubner, Stuttgart, 2003.

[17] K. HAMAGUCHI, *The protein molecule*, Japan Scientific Societies Press, Tokyo, 1992.

[18] I. N. LEVINE, *Quantum Chemistry*, Prentice Hall, New Jersey, 2000.

[19] P. W. ATKINS, *Quanta: A Handbook of Concepts*, Oxford University Press, Oxford, 1991.

[20] A. R. LEACH, *Molecular modelling: principles and applications*, Prentice Hall, Harlow, 2001.

[21] M. HERZ, F. J. GIESSIBL, AND J. MANNHART, *Probing the shape of atoms in real space*, Phys. Rev. B, 68 (2003), pp. 045301-1–045301-7.

[22] R. S. MULLIKEN, *A new electroaffinity scale; together with data on valence states and on valence ionization potentials and electron affinities*, J. Chem. Phys., 2 (1934), pp. 782–793.

[23] J. HINZE AND H. H. JAFFÉ, *Electronegativity. I. Orbital electronegativity of neutral atoms.*, J. Amer. Chem. Soc., 84 (1962), pp. 540–546.

[24] J. HINZE, M. A. WHITEHEAD, AND H. H. JAFFÉ, *Electronegativity. II. Bond and orbital electronegativities*, J. Amer. Chem. Soc., 85 (1963), pp. 148–154.

[25] J. HINZE AND H. H. JAFFÉ, *Electronegativity. IV. Orbital electronegativities of the neutral atoms of the periods three A and four A and of positive ions of periods one and two*, J. Phys. Chem., 67 (1963), pp. 1501–1506.

[26] J. GASTEIGER AND M. MARSILI, *A new model for calculating atomic charges in molecules*, Tetrahedron Letters, 34 (1978), pp. 3181–3184.

[27] ——, *Iterative partial equalization of orbital electronegativity - a rapid access to atomic charges*, Tetrahedron, 36 (1980), pp. 3219–3288.

[28] E. BLANC, V. FREMONT, P. SIZUN, S. MEUNIER, J. V. RIETSCHOTEN, A. THEVAND, J.-M. BERNASSAU, AND H. DARBON, *Solution structure of P01, a natural scorpion peptide structurally analogous to scorpion toxins specific for apamin-sensitive potassium channel*, Proteins: Struct., Funct., Genet., 24 (1996), pp. 359–369.

[29] A. C. LEGON AND D. J. MILLEN, *Directional character, strength, nature of the hydrogen bond in gas-phase dimers*, Acc. Chem. Res., 20 (1987), pp. 39–46.

[30] L. VITAGLIANO, R. BERISIO, A. MASTRANGELO, L. MAZZARELLA, AND A. ZAGARI, *Preferred proline puckerings in cis and trans peptide groups: Implications for collagen stability*, Protein Sci., 10 (2001), pp. 2627–2632.

[31] A. M. KREZEL, C. KASIBHATLA, P. HIDALGO, R. MACKINNON, AND G. WAGNER, *Solution structure of the potassium channel inhibitor agitoxin 2: caliper for probing channel geometry*, Protein Sci., 4 (1995), pp. 1478–1489.

[32] M. MIZUGUCHI, S. KAMATA, S. KAWABATA, N. FUJITANI, AND K. KAWANO, *Solution structure of tachyplesin I in dodecylphosphocholine*, RCSB Protein Data Bank record 1WO1, (2005).

[33] R. E. DICKERSON AND I. GEIS, *The structure and action of proteins*, Harper & Row Publishers, New York, 1969.

[34] B. I. LEE AND S. W. SUH, *Crystal structure of the schiff base intermediate prior to decarboxylation in the catalytic cycle of aspartate α-decarboxylase*, J. Mol. Biol., 340 (2004), pp. 1–7.

[35] A. SEYEDARABI, T. T. TO, S. ALI, S. HUSSAIN, M. FRIES, R. MADSEN, M. H. CLAUSEN, S. TEIXTEIRA, K. BROCKLEHURST, AND R. W. PICKERSGILL, *Structural insights into substrate specificity and the anti β-elimination mechanism of pectate lyase*, Biochemistry, 49 (2010), pp. 539–540.

[36] P. Y. CHOU AND G. D. FASMAN, *Prediction of protein conformation*, Biochemistry, 13 (1974), pp. 222–245.

[37] G. FERMI, M. F. PERUTZ, B. SHAANAN, AND R. FOURME, *The crystal structure of human deoxyhaemoglobin at 1.74 Å resolution*, J. Mol. Biol., 175 (1984), pp. 159–174.

[38] K. M. BISWAS, D. R. DEVIDO, AND J. G. DORSEY, *Evaluation of methods for measuring amino acid hydrophobicities and interactions*, J. Chromatogr. A, 1000 (2003), pp. 637–655.

[39] Y. NOZAKI AND C. TANFORD, *The solubility of amino acids and two glycine peptides in aqueous ethanol and dioxane solutions. Establishment of a hydrophobicity scale*, J. Biol. Chem., 246 (1971), pp. 2211–2217.

[40] M. J. E. STERNBERG, *Protein structure prediction*, Oxford University Press, New York, 1996.

[41] E. Q. LAWSON, A. J. SADLER, D. HARMATZ, D. T. BRANDAU, R. MICANOVIC, R. D. MACELROY, AND C. R. MIDDAUGH, *A simple experimental model for hydrophobic interactions in proteins*, J. Biol. Chem., 259 (1984), pp. 2910–2912.

[42] B. LEE AND F. M. RICHARDS, *The interpretation of protein structures: estimation of static accessibility*, J. Mol. Biol., 55 (1971), pp. 379–400.

[43] A. S. SPIRIN, *Ribosomes*, Kluwer Academic/Plenum Publishers, New York, 1999.

[44] A. R. KUSMIERCZYK AND J. MARTIN, *Chaperonins - keeping a lid on folding proteins*, FEBS Lett., 505 (2001), pp. 343–347.

[45] M. P. MORRISSEY, Z. AHMED, AND E. I. SHAKHNOVICH, *The role of co-translation in protein folding: a lattice model study*, Polymer, 45 (2004), pp. 557–571.

[46] V. A. KOLB, E. V. MAKEYEV, AND A. S. SPIRIN, *Folding of firefly luciferase during translation in a cell-free system*, EMBO J., 13 (1994), pp. 3631–3637.

[47] A. N. FEDOROV AND T. O. BALDWIN, *Contribution of cotranslational folding to the rate of formation of native protein structure*, Proc. Natl. Acad. Sci. USA, 92 (1995), pp. 1227–1231.

[48] R. W. RUDDON AND E. BEDOWS, *Assisted protein folding*, J. Biol. Chem., 272 (1997), pp. 3125–3128.

[49] G. KRAMER, V. RAMACHANDIRAN, AND B. HARDESTY, *Cotranslational folding — omnia mea mecum porto?*, Int. J. Biochem. Cell Biol., 33 (2001), pp. 541–553.

[50] R. J. ELLIS, *Chaperone substrates inside the cell*, Trends Biochem. Sci., 25 (2000), pp. 210–212.

[51] M. SHTILERMAN, G. H. LORIMER, AND S. W. ENGLANDER, *Chaperonin function: folding by forced unfolding*, Science, 284 (1999), pp. 822–825.

[52] K.-M. PAN, M. BALDWIN, J. NGUYEN, M. GASSET, A. SERBAN, D. GROTH, I. MEHLHORN, Z. HUANG, R. J. FLETTERICK, F. E. COHEN, AND S. B. PRUSINER, *Conversion of α-helices into β-sheets features in the formation of the scrapie prion proteins*, Proc. Natl. Acad. Sci. USA, 90 (1993), pp. 10962–10966.

[53] R. ZAHN, *Prion propagation and molecular chaperones*, Q. Rev. Biophys., 32 (1999), pp. 309–370.

[54] D. L. NELSON AND M. M. COX, *Lehninger Principles of Biochemistry*, W. H. Freeman and Company, New York, 2008.

[55] T. TENSON AND M. EHRENBERG, *Regulatory nascent peptides in the ribosomal tunnel*, Cell, 108 (2002), pp. 591–594.

[56] C. BERNABEU AND J. A. LAKE, *Nascent polypeptide chains emerge from the exit domain of the large ribosomal subunit: immune mapping of the nascent chain*, Proc. Natl. Acad. Sci. USA, 79 (1982), pp. 3111–3115.

[57] D. G. MORGAN, J.-F. MÉNÉTRET, M. RADERMACHER, A. NEUHOF, I. V. AKEY, T. A. RAPOPORT, AND C. W. AKEY, *A comparison of the yeast and rabbit 80 S ribosome reveals the topology of the nascent chain exit tunnel, intersubunit bridges and mammalian rRNA expansion segments*, J. Mol. Biol., 301 (2000), pp. 301–321.

[58] R. BECKMANN, D. BUBECK, R. GRASSUCCI, P. PENCZEK, A. VERSCHOOR, G. BLOBEL, AND J. FRANK, *Alignment of conduits for the nascent polypeptide chain in the ribosome-Sec61 complex*, Science, 278 (1997), pp. 2123–2126.

[59] R. BECKMANN, C. M. T. SPAHN, N. ESWAR, J. HELMERS, P. A. PENCZEK, A. SALI, J. FRANK, AND G. BLOBEL, *Architecture of the protein-conducting channel associated with the translating 80S ribosome*, Cell, 107 (2001), pp. 361–372.

[60] T. TENSON, M. LOVMAR, AND M. EHRENBERG, *The mechanism of action of macrolides, lincosamides and streptogramin B reveals the nascent peptide exit path in the ribosome*, J. Mol. Biol., 330 (2003), pp. 1005–1014.

[61] F. GONG AND C. YANOFSKY, *Instruction of translating ribosome by nascent peptide*, Science, 297 (2002), pp. 1864–1867.

[62] R. BERISIO, F. SCHLUENZEN, J. HARMS, A. BASHAN, T. AUERBACH, D. BARAM, AND A. YONATH, *Structural insight into the role of the ribosomal tunnel in cellular regulation*, Nat. Struct. Biol., 10 (2003), pp. 366–370.

[63] J. POEHLSGAARD AND S. DOUTHWAITE, *The bacterial ribosome as a target for antibiotics*, Nat. Rev. Microbiol., 3 (2005), pp. 870–881.

[64] C. M. SPAHN, M. G. GOMEZ-LORENZO, R. A. GRASSUCCI, R. JORGENSEN, G. R. ANDERSEN, R. BECKMANN, P. A. PENCZEK, J. P. BALLESTA, AND J. FRANK, *Domain movements of elongation factor eEF2 and the eukaryotic 80S ribosome facilitate tRNA translocation*, Embo J., 23 (2004), pp. 1008–1019.

[65] W. D. PICKING, O. W. ODOM, T. TSALKOVA, I. SERDYUK, AND B. HARD-ESTY, *The conformation of nascent polylysine and polyphenylalanine peptides on ribosomes*, J. Biol. Chem., 266 (1991), pp. 1534–1542.

[66] L. A. RYABOVA, O. M. SELIVANOVA, V. I. BARANOV, V. D. VASILIEV, AND A. S. SPIRIN, *Does the channel for nascent peptide exist inside the ribosome? Immune electron microscopy study*, FEBS Lett., 226 (1988), pp. 255–260.

[67] V. RAMACHANDIRAN, C. WILLMS, G. KRAMER, AND B. HARDESTY, *Fluorophores at the N terminus of nascent chloramphenicol acetyltransferase peptides affect translation and movement through the ribosome*, J. Biol. Chem., 275 (2000), pp. 1781–1786.

[68] E. WESTHOF, P. DUMAS, AND D. MORAS, *Restrained refinement of two crystalline forms of yeast aspartic acid and phenylalanine transfer RNA crystals*, Acta Cryst.,Sect.A, 44 (1988), pp. 112–123.

[69] H.-J. RHEINBERGER, H. STERNBACH, AND K. H. NIERHAUS, *Codon-anticodon interaction at the ribosomal E site*, J. Biol. Chem., 261 (1986), pp. 9140–9143.

[70] L. JENNER, B. REES, M. YUSUPOV, AND G. YUSUPOVA, *Messenger RNA conformations in the ribosomal E site revealed by X-ray crystallography*, EMBO reports, 8 (2007), pp. 846–850.

[71] V. MÁRQUEZ, D. N. WILSON, W. P. TATE, F. TRIANA-ALONSO, AND K. H. NIERHAUS, *Maintaining the ribosomal reading frame: the influence of the E site during translational regulation of release factor 2*, Cell, 118 (2004), pp. 45–55.

[72] V. I. LIM AND A. S. SPIRIN, *Stereochemical analysis of ribosomal transpeptidation. Conformation of nascent peptide*, J. Mol. Biol., 188 (1986), pp. 565–577.

[73] A. BASHAN, I. AGMON, R. ZARIVACH, F. SCHLUENZEN, J. HARMS, R. BERISIO, H. BARTELS, F. FRANCESCHI, T. AUERBACH, H. A. S. HANSEN, E. KOSSOY, M. KESSLER, AND A. YONATH, *Structural basis of the ribosomal machinery for peptide bond formation, translocation, and nascent chain progression for peptide bond formation, translocation, and nascent chain progression*, Mol. Cell, 11 (2003), pp. 91–102.

[74] I. AGMON, T. AUERBACH, D. BARAM, H. BARTELS, A. BASHAN, R. BERISIO, P. FUCINI, H. A. S. HANSEN, J. HARMS, M. KESSLER, M. PERETZ, F. SCHLUENZEN, A. YONATH, AND R. ZARIVACH, *On peptide bond formation, translocation, nascent protein progression and the regulatory properties of ribosomes*, Eur. J. Biochem., 270 (2003), pp. 2543–2556.

[75] A. BASHAN AND A. YONATH, *Ribosome crystallography: catalysis and evolution of peptide-bond formation, nascent chain elongation and its co-translational folding*, Biochem. Soc. Trans., 33 (2005), pp. 488–492.

[76] S. PAL, S. CHANDRA, S. CHOWDHURY, D. SARKARI, A. N. GHOSH, AND C. DASGUPTA, *Complementary role of two fragments of domain V of 23 S ribosomal RNA in protein folding*, J. Biol. Chem., 274 (1999), pp. 32771–32777.

[77] S. CHOWDHURY, S. PAL, J. GHOSH, AND C. DASGUPTA, *Mutations in domain V of the 23S ribosomal RNA of Bacillus subtilis that inactivate its protein folding property in vitro*, Nucleic Acids Res., 30 (2002), pp. 1278–1285.

[78] D. SAMANTA, D. MUKHOPADHYAY, S. CHOWDHURY, J. GHOSH, S. PAL, A. BASU, A. BHATTACHARYA, A. DAS, D. DAS, AND C. DASGUPTA, *Protein folding by domain V of Escherichia coli 23S rRNA: specificity of RNA-protein interactions*, J. Bacteriol., 190 (2008), pp. 3344–3352.

[79] W. KUDLICKI, A. COFFMAN, G. KRAMER, AND B. HARDESTY, *Ribosomes and ribosomal RNA as chaperones for folding of proteins*, Folding & Design, 2 (1997), pp. 101–108.

[80] S. C. SANYAL, S. PAL, S. CHAUDHURI, AND C. DASRUPTA, *23S rRNA assisted folding of cytoplasmic malate dehydrogenase is distinctly different from its self-folding*, Nucleic Acids Res., 30 (2002), pp. 2390–2397.

[81] L. I. MALKIN AND A. RICH, *Partial resistance of nascent polypeptide chains to proteolytic digestion due to ribosomal shelding*, J. Mol. Biol., 26 (1967), pp. 329–346.

[82] G. BLOBEL AND D. D. SABATINI, *Controlled proteolysis of nascent polypeptides in rat liver cell fractions*, J. Cell Biol., 45 (1970), pp. 130–145.

[83] S. A. ETCHELLS AND F. U. HARTL, *The dynamic tunnel*, Nat. Struct. Mol. Biol., 11 (2004), pp. 391–392.

[84] T. TSALKOVA, O. W. ODOM, G. KRAMER, AND B. HARDESTY, *Different conformations of nascent peptides on ribosomes*, J. Mol. Biol., 278 (1998), pp. 713–723.

[85] C. A. WOOLHEAD, P. J. MCCORMICK, AND A. E. JOHNSON, *Nascent membrane and secretory proteins differ in FRET-detected folding far inside the ribosome and in their exposure to ribosomal proteins*, Cell, 116 (2004), pp. 725–736.

[86] W. H. ELLIOTT AND D. C. ELLIOTT, *Biochemistry and molecular biology (Russian translation)*, MAIK "Nauka/Interperiodika", Moskow, 2002.

[87] N. C. COHEN, ED., *Guidebook on molecular modeling in drug design*, Academic Press, San Diego, 1996.

[88] G. LIPPENS, J. NAJIB, S. J. WODAK, AND A. TARTAR, *NMR sequential assignments and solution structure of chlorotoxin, a small scorpion toxin that blocks chloride channels*, Biochemistry, 34 (1995), pp. 13–21.

[89] D. COZZETTO, A. KRYSHTAFOVYCH, K. FIDELIS, J. MOULT, B. ROST, AND A. TRAMONTANO, *Evaluation of template-based models in casp8 with standard measures*, Proteins, 77 (2009), pp. 18–28.

[90] J. B. STURGEON AND B. B. LAIRD, *Symplectic algorithm for constant-pressure molecular dynamics using a Nosé-Poincaré thermostat*, J. Chem. Phys., 112 (2000), pp. 3474–3482.

[91] B. LEIMKUHLER, C. CHIPOT, R. ELBER, A. LAAKSONEN, A. MARK, T. SCHLICK, C. SCHÜTTE, AND R. SKEEL, EDS., *New algorithms for macromolecular simulations*, Springer, Berlin, 2006.

[92] V. A. VOELZ, G. R. BOWMAN, K. BEAUCHAMP, AND V. S. PANDE, *Molecular simulation of ab initio protein folding for a millisecond folder NTL9(1-39)*, J. Am. Chem. Soc., 132 (2010), pp. 1526–1528.

[93] J. GARNIER, D. J. OSGUTHORPE, AND B. ROBSON, *Analysis of the accuracy and implications of simple methods for predicting the secondary structure of globular proteins*, J. Mol. Biol., 120 (1978), pp. 97–120.

[94] M. CLARK, R. D. C. III, AND N. VAN OPDENBOSCH, *Validation of the general purpose Tripos 5.2 force field*, J. Comput. Chem., 10 (1989), pp. 982–1012.

[95] T. A. HALGREN, *Merck molecular force field. I. Basis, form, scope, parametrization, and performance of MMFF94*, J. Comput. Chem., 17 (1996), pp. 490–519.

[96] X. DAURA, A. E. MARK, AND W. F. VAN GUNSTEREN, *Parametrization of aliphatic CH_n united atoms of GROMOS96 force field*, J. Comput. Chem., 19 (1998), pp. 535–547.

[97] M. L. P. PRICE, D. OSTROVSKY, AND W. L. JORGENSEN, *Gas-phase and liquid-state properties of esters, nitriles, and nitro compounds with the OPLS-AA force field*, J. Comput. Chem., 22 (2001), pp. 1340–1352.

[98] J. WANG, R. M. WOLF, J. W. CARDWELL, P. A. KOLLMAN, AND D. A. CASE, *Development and testing of a general Amber force field*, J. Comput. Chem., 25 (2004), pp. 1157–1174.

[99] A. HINCHLIFFE, *Molecular modelling for beginners*, Wiley, Chichester, 2003.

[100] W. L. JORGENSEN AND J. TIRADO-RIVES, *Potential energy functions for atomic-level simulations of water and organic and biomolecular systems*, Proc. Natl. Acad. Sci USA, 102 (2005), pp. 6665–6670.

[101] D. E. WILLIAMS, *Improved intermolecular force field for molecules containing H, C, N, and O atoms, with application to nucleoside and peptide crystals*, J. Comput. Chem., 22 (2001), pp. 1154–1166.

[102] U. C. SINGH AND P. A. KOLLMAN, *An approach to computing electrostatic charges for molecules*, J. Comput. Chem., 5 (1984), pp. 129–145.

[103] A. L. MCCLELLAN, *Tables of experimental dipole moments*, W. H. Freeman and Company, San Francisco and London, 1963.

[104] F. A. MOMANY, R. F. MCGUIRE, A. W. BURGESS, AND H. A. SCHERAGA, *Energy parameters in polypeptides. VII. Geometric parameters, partial atomic charges, nonbonded interactions, hydrogen bond interactions, and intrinsic torsional potentials for the naturally occuring amino acids*, J. Phys. Chem., 79 (1975), pp. 2361–2381.

[105] S. J. WEINER, P. A. KOLLMAN, D. A. CASE, U. C. SINGH, C. GHIO, G. ALAGONA, S. PROFETA, JR., AND P. WEINER, *A new force field for molecular mechanical simulation of nucleic acids and proteins*, J. Am. Chem. Soc., 106 (1984), pp. 765–784.

[106] J.-H. LII AND N. L. ALLINGER, *Directional hydrogen bonding in the MM3 force field. I*, J. Phys. Org. Chem., 7 (1994), pp. 591–609.

[107] ——, *Directional hydrogen bonding in the MM3 force field. II*, J. Comput. Chem., 19 (1998), pp. 1001–1016.

[108] A. SHRAKE AND J. A. RUPLEY, *Environment and exposure to solvent of protein Atoms. Lysozyme and insulin*, J. Mol. Biol., 79 (1973), pp. 351–371.

[109] M. L. CONNOLLY, *Analytical molecular surface calculation*, J. Appl. Cryst., 16 (1983), pp. 548–558.

[110] T. J. RICHMOND, *Solvent accessible surface area and excluded volume in proteins*, J. Mol. Biol., 178 (1984), pp. 63–89.

[111] N. FUTAMURA, S. ALURU, D. RANJAN, AND B. HARIHARAN, *Efficient parallel algorithms for solvent accessible surface area of proteins*, IEEE Trans. on Paral. and Distr. Sys., 13 (2002), pp. 544–555.

[112] S. J. WODAK AND J. JANIN, *Analytical approximation to the accessible surface area of proteins*, Proc. Natl. Acad. Sci. USA, 77 (1980), pp. 1736–1740.

[113] W. HASEL, T. F. HENDRICKSON, AND W. C. STILL, *A rapid approximation to the solvent accessible surface areas of atoms*, Tetrahedron Comput. Methodol., 1 (1988), pp. 103–116.

[114] L. CAVALLO, J. KLEINJUNG, AND F. FRATERNALI, *POPS: a fast algorithm for solvent accessible surface areas at atomic and residue level*, Nucleic Acids Res., 31 (2003), pp. 3364–3366.

[115] F. M. RICHARDS, *Areas, volumes, packing and protein structure*, Ann. Rev. Biophys. Bioeng., 6 (1977), pp. 151–176.

[116] B. MALLIK, A. MASUNOV, AND T. LAZARIDIS, *Distance and exposure dependent effective dielectric function*, J. Comput. Chem., 23 (2002), pp. 1090–1099.

[117] B. CASSELMAN, *The difficulties of kissing in three dimensions*, Notices Amer. Math. Soc., 51 (2004), pp. 884–885.

[118] K. SCHÜTTE AND B. L. VAN DER WAERDEN, *Das Problem der dreizehn Kugeln*, Math. Annalen, 125 (1953), pp. 325–334.

[119] N. COPERNICUS, *De revolutionibus orbium coelestium*, Norimbergae: apud loh. Petreium, 1543.

[120] I. N. BRONSTEIN, K. A. SEMENDJAJEW, G. MUSIOL, AND H. MÜHLIG, *Taschenbuch der Mathematik*, Harri Deutsch, Thun und Frankfurt am Main, 2001.

[121] M. WOO, J. NEIDER, T. DAVIS, AND D. SHREINER, *OpenGL programming guide*, Addison-Wesley, Boston, 2002.

[122] L. D. LANDAU AND E. M. LIFSHITZ, *Mechanics (Russian edition)*, Fizmatlit, Moscow, 2002.

[123] S. VIJAY-KUMAR, C. E. BUGG, AND W. J. COOK, *Structure of ubiquitin refined at 1.8 Å resolution*, J. Mol. Biol., 194 (1987), pp. 531–544.

[124] A. BONDI, *van der Waals volumes and radii*, J. Phys. Chem., 68 (1964), pp. 441–451.

CURRICULUM VITAE

Personal Data:

Name: Anna Shumilina

Date of Birth: 07.09.1977

Place of Birth: Moscow, Russia

School Education:

1984 – 1990   Secondary School #2,
                  Vladivostok, Russia

1986 – 1991   Art department of Musical School #3,
                  Vladivostok, Russia

1990 – 1994   Mathematical Secondary School #23,
                  Vladivostok, Russia

1991 – 1994   Young Academy of Marine Biology,
                  Vladivostok, Russia

University Education:

1994 – 1999   Department of Mathematics and Computer science,
                  Far Eastern National University, Vladivostok, Russia

1994 – 1999   Department of Biology, Ecology and Soil Science,
                  Far Eastern National University, Vladivostok, Russia

1999 – 2001   Department of Mathematics,
                  University of Kaiserslautern, Germany

September 2001   Master of Science in Industrial Mathematics

2007 – 2009   PhD Fellowship from Fraunhofer-Kolleg,
                  TU Kaiserslautern / Fraunhofer ITWM, Germany

Academic Career:

2001 – 2007   Department of Mathematics,
                  Technical University of Berlin, Germany

2009 – to date   Fraunhofer ITWM, Kaiserslautern, Germany

Lebenslauf

Persönliche Daten:

Name:	Anna Shumilina
Geburtsdatum:	07.09.1977
Geburtsort:	Moskau, Russland

Schulbildung:

1984 – 1990	Mittelschule #2, Vladivostok, Russland
1986 – 1991	Kunstabteilung der Musikschule #3, Vladivostok, Russland
1990 – 1994	Mittelschule #23, Vladivostok, Russland
1991 – 1994	Junge Akademie für Marinebiologie, Vladivostok, Russland

Universitätsstudium:

1994 – 1999	Fachbereich für Mathematik und Informatik, Fernöstliche Staatsuniversität, Vladivostok, Russland
1994 – 1999	Fachbereich für Biologie, Ökologie und Bodenkunde, Fernöstliche Staatsuniversität, Vladivostok, Russland
1999 – 2001	Fachbereich für Mathematik, Universität Kaiserslautern, Deutschland
September 2001	Master of Science in Technomathematik
2007 – 2009	Doktorandenstipendium von Fraunhofer-Kolleg, TU Kaiserslautern / Fraunhofer ITWM, Deutschland

Berufstätigkeit:

2001 – 2007	Fachbereich für Mathematik, Technische Universität Berlin, Deutschland
2009 – bis jetzt	Fraunhofer ITWM, Kaiserslautern, Deutschland